A Journey Through the Realm
of Theoretical Chemistry

A Journey Through the Realm of Theoretical Chemistry

Fereydoon Milani-nejad Ph.D.

Copyright © 2015 by Fereydoon Milani-nejad Ph.D.

ISBN:	Hardcover	978-1-5144-7179-1
	Softcover	978-1-5144-1101-8
	eBook	978-1-5144-1100-1

All rights reserved. No part of this book may be reproduced or transmitted in any form or by any means, electronic or mechanical, including photocopying, recording, or by any information storage and retrieval system, without permission in writing from the copyright owner.

Any people depicted in stock imagery provided by Thinkstock are models, and such images are being used for illustrative purposes only.
Certain stock imagery © Thinkstock.

Print information available on the last page.

Rev. date: 02/27/2016

To order additional copies of this book, contact:
Xlibris
1-888-795-4274
www.Xlibris.com
Orders@Xlibris.com
725939

CONTENTS

Preface ix

Chapter One: Overview 1

Chapter Two: Microscopic Description of Matter 14

Chapter Two Supplementary 26

Chapter Three: Macroscopic Description of Matter Behavior 32

Chapter Three Supplementary 52

Chapter Four: Reactions 64

Chapter Five: Applications 76

Fereydoon Milani-nejad Ph.D.

Professor of Emeritus

Department of Chemistry

Ferdowsi University

Meshad, Iran

Current Address

4 Belle Isle Dr.

Laguna Niguel

CA, 92677

femilani@aol.com

Preface

The primary goal of this book is to show that theoretical chemistry is not just a topic that ought to be explored by science majors. In fact, I aim to present theoretical chemistry as a discipline that is applicable to many other fields including sociology, psychology, philosophy, the social sciences, and even theology. I've devoted Chapter One of this book, the "Overview", chiefly to this task. In this chapter, by establishing a qualitative correlation between human behavior and molecular behavior, I elucidate on how the enlightenment of the mind, fate and will, partnership (marriage) and its stability, dual personalities, excited behaviors, revelation, excited universes (parallel universes), and more can all be explained in terms of chemical laws.

The concept of potential well, qualitatively introduced in Chapter One, is explored quantitatively in Chapter Two by means of a very simple example, while methods for extracting information from potential well at microscopic level are also discussed. To keep discussion as simple as possible, this chapter and chapter three are divided into main and supplementary sections. In the main section of each chapter the subject matter is presented in a qualitative manner. In the supplementary sections, the subject is discussed at a moderate level, avoiding rigorous treatment of the subject matter. In both chapters application of the laws to behavior are outlined.

Chapter three emphasizes on the equality, justice concept, and evolutionary concept of entropy rather than other concepts such as disorder, randomness, information, gradual decline, uncertainty, and more.

Chapter Four discusses reactions from both macroscopic and microscopic views using a simple reaction as an example.

Chapter Five is devoted to the application of the theories discussed in Chapters Two, Three and Four to simple atoms such as hydrogen and oxygen, and to simple molecules such as water.

Chapter One

Overview

Chemistry is a subject whose boundaries stretch as far as the boundaries of the universe. Correspondingly it is as mysterious and attractive as the universe itself, yet to date, chemistry has yet to be widely appreciated as much as music, movies, poetry, drama, etc. Ironically enough, appreciations or disapproval processes happens to be controlled by the laws of chemistry. In principal this statement is justified on the ground that these processes are activated, controlled, terminated, and recorded by some kind of chemical reactions in the mind.

The mind-body relation, that is, the ability of the mind to act upon matter and the effect of matter upon the mind, is subject of interest to psychologists, philosophers, social scientists, and scientists alike. In the past decades, interdisciplinary majors in biochemistry and psychology have been setup in many universities to explore the relation between the physico-chemical behaviors of matter and human behavior. The concepts of quantum mind and quantum consciousness has attracted the attention of sociologist, philosophers, and scientists.

Chemical reactions are controlled by chemical laws, therefore by becoming familiar with the general concepts of theoretical chemistry and the well-established laws that govern chemical behavior of matter, one will be able to understand human behavior at least in a qualitative manner, and thus study chemistry with more interest.

In order to demonstrate that the laws governing human behavior and molecular behavior are the same, we need to show that at least in a qualitative manner there exists a one to one correlation between the two. With this aim in mind, we will view human behavior and molecular behavior through the same window equipped with a special kind of lens. For this purpose we define the reduced distance "R_{red}", a dimensionless quantity as the ratio of distance between two objects "r" to the mean radius of the objects "$r°$". As a specific example, consider hydrogen atoms at molecular level. The hydrogen atom is described as a hydrogen nucleus (proton) with an associated electron that may be found with a maximum probability in spherical shell of 0.53 Å (1 Å =10^{-10} m). For simplicity we assume that the radius of the hydrogen atom roughly to be about 10^{-10} m. For two hydrogen atoms one millimeter apart, the reduced distance "R_{red}" will be 10^7. Now consider two individual people, then for the same reduced distance these two should be 10^7 m or 10^4 km apart. This distance between the two is comparable with the separation of one millimeter between two hydrogen atoms. At this distance there is no feeling, no interaction between these people, and also there exists no interaction between the two hydrogen atoms. As two hydrogen atoms are brought closer to each other, each atom will feel the presence of the other and if the two atoms are in the bonding state (right mood) then the interaction will be attractive. Correspondingly, if they are in none bonding state (wrong mood) then the interaction will be repulsive. In analogy, if the two people are in the right mood, an attractive interaction builds up between the two as they approach each other. For two hydrogen atoms the attractive interaction between the two has its maximum value at 0.74 Å, and a stable system (molecule) is formed. Beyond this point the attractive interaction rapidly decreases and becomes repulsive. The variation of interaction energy with distance as shown in Figure 1-1 is called potential energy well.

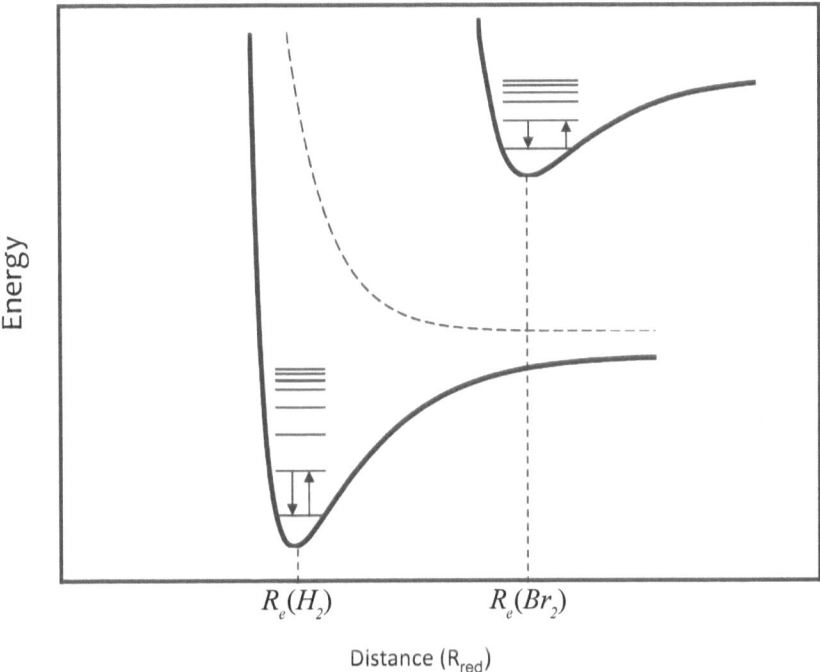

Figure 1-1 Schematic representation of the interaction between two bodies as a function of reduced distance (R_{red}) between the two. The solid line represents an attractive interaction (the potential energy well). The dashed curve represents a repulsive interaction. The horizontal lines inside the well represent the energy levels (stages) available for occupation. The spacing of energy stages depends on the shape and depth of the potential well. The potential energy well for a larger and heavier body is shown on the upper portion of the figure.

The hydrogen atoms in hydrogen molecule have a vibratory motion with respect to each other. There exists discrete energy levels associated with this motion and they are shown by horizontal lines in the potential energy wells in Figure 1-1. There are also other energy levels associated with rotational motion of the molecule, but they are not shown in this Figure[1]. The spacing between these energy levels depends on the shape and depth of the potential energy well. Quantum chemistry calculations show that the separation of these energy levels, that is, the depth and width of the potential energy well, depend on the properties of the constituent atoms and their arrangements (geometry). As the atoms become heavier and larger, the spacing between energy levels reduces. The spacing of energy levels, that is, the shape and depth of this potential energy well determine all microscopic and macroscopic properties of the molecule.

It is proposed that there exists a behavioral well for interaction between people. The depth and shape of the behavioral well determines human behavior, and information about the behavior can be deduced from the spacing between levels (stages) in the behavioral well.

For simplicity, the concept of the potential well was introduced for the interaction of two simple atoms to form a simple molecule. But, atoms are made of electrons, protons, and neutrons. A similar potential well describes the properties of that atom. The same holds for the interaction of elementary particles. In analogy, for each person there exists a behavior well where the depth, shape, and spacing between stages within the well describe the behavior of that person.

---

[1] There are 3N degrees of freedom for molecular motion, N being the number of atoms in the molecule. Three degrees of freedom are associated with the translational motion of the molecule in space. In the case of linear molecules, two degrees of freedom are associated with internal rotation and 3N-5 with vibrational motion. In the case of none linear molecules there are three degrees of freedom for rotation and 3N-6 degrees of freedom for vibrational motion.

We will discuss both microscopic and macroscopic properties of matter through this window "the potential well", and will show that the shape and depth of this potential well and spacing of energy levels determines all microscopic properties of the molecule and the kinetics of reaction between molecules. We are also proposing that the depth and shape of the potential well for interaction between people "behavior well", and spacing of the stages within the well determine human behavior. Mostly in this chapter and occasionally in other chapters through this book we will correlate the physico-chemical properties of matter with human behavior and through this correlation we will try to demonstrate that the chemical laws apply to human behavior, or in another words both obey the same laws.

Imagine that many molecules of the same type (hydrogen molecules in this example) are confined into a vessel under a given pressure and temperature. At this point we will neglect the effect of collision pressure on the shape of the potential well, that is, we assume an ideal gas behavior for the individual molecules.

In none-elastic collision between molecules, if some specific criteria are met, kinetic energy will transfer from one molecule to the internal modes of the other[2]. Then, for a large number of molecules (an ensemble of molecules), the molecules will be distributed among the energy levels according to Maxwell- Boltzmann's distribution law. This distribution depends on the temperature and separation of energy levels. Recall that the latter depends on the shape of the potential energy well and the former depends on the environment. This means that as the temperature increases the number of collisions will increase and the upper energy levels will become more populated, and if the temperature is high enough (the effect of environment) the molecules will fall apart.

---

[2] In none elastic collision the kinetic energy will be transferred from one body to the internal modes of energy (Energy states) of the other molecule. In elastic collision only the kinetic energy will be transferred from one body to the other.

The containment of hydrogen molecules in a vessel is almost similar to people living in an isolated village. The behavior well for each individual is different from the other, but almost similar. For these people their fate is written on their foreheads, that is, the depth and the shape of the behavior well for them approximately could be predicted by their past history. The energy stage that individual occupies (The Will) depends on the energy and activity of that individual.

Interaction of light and matter (spectroscopy) temporarily will transfer molecules from one energy level to another energy level. The energy transfer will occur if the energy of light packets is equal to the energy difference between two energy levels, and such a transition is allowed (determined by the properties of the two energy levels involved). This type of interaction provides information about the potential well.

Since the behavior well for a person (a couple, a society) is unique with unique energy stages, then a vision, a sensation, and even a dream could enlighten the mind to transfer from one stage to another. By analogy, in molecular case, it is expected that transition will take place if the energy provided by such process is exactly equal to the difference of energy between the two stages, and if the transition between the two stages is allowed. The process is termed enlightenment of the soul. Revelation produces information about the stages and the potential well describing behavior.

As shown in figure 1-1, in order to bring the two atoms together closer than r_e, a tremendous amount of energy is needed. Fusion of the two deuterium atoms will produce a helium atom and releases a tremendous amount of energy. Experimentally, it is impossible to fuse two people and produce a new person, but uplifting of the soul through hardship has been the subject of many philosophical articles[3].

If the two hydrogen atoms are not in the right mood, that is the electrons in the two approaching atoms have parallel spins (spin is a quantum mechanical property of the matter), then the interaction will be repulsive (dashed curve in figure 1-1), leading to an anti-bonding state. A pulse of energy can transfer the system from its bonding state to nonbonding state. If the system does not cool off immediately and return to its equilibrium state, then the two bodies will fall apart.

For an "Earthly" partnership (the term "Earthly" here is correlated with the ground electronic state of the matter) each member should have at least one "state of mind" (the term "state of mind" is correlated with the spin function of the particle) different from others in order for the interaction to be attractive. We have referred to this requirement as the "right mood". For parallel "state of mind" the interaction will be repulsive. A pulse of temper will transfer the bonding state to an anti-bonding state and if the temperament is not quenched properly, then the process will lead to separation.

For larger and heavier atoms, for example two bromine atoms, the potential well is broader, the distance between two bodies at minimum "r_e" is longer, the well is shallower, and the energy levels are closer to each other. Therefore, at the same temperature the upper levels for these types of molecules are more populated than the

---

[3] As an example see the Conference of birds by Iranian philosopher and poet, Farid od-Din Attar (c 1200). This allegorical poem describes the quest of birds for the mystical Simorgh, or Phoenix, whom they wish to be their king (i.e., God). Many birds make excuses, for they do not wish to continue, and the excuse symbolizes those made by men for not pursing spiritual perfection. Of those who had begun the pilgrimage only 30 birds (si morgh) succeed in entering the presence of the Simorgh. In the final scene the 30 birds approach the throne contemplating their reflections in the mirror like countenance of the Simorgh, only to realize that they (si morgh) and the Simorgh are one.

previous pairs, and as a result the molecule will be less stable. When the bond between two atoms in this molecule is broken the result will be two reactive radicals ready to react with other bodies that they collide with, for example with hydrogen molecule to form hydrogen bromide with a characteristic potential well of its own.

When the potential well is deep and narrow, then the tendency for association with others decreases, for example a hydrogen molecule does not interact with other hydrogen molecules and therefore at ordinary and even low temperatures it has a gaseous appearance. As the depth of the well decreases and it becomes broader, the tendency for association with other molecules increases. Thus Br_2 associates with other Br_2 molecules and at ordinary temperature it is a liquid. Such behavior is observed for families and societies, that is, when the relation between the members of a family or relations within a society is strong (deep behavioral well) then the tendency to associate with other families or other societies is less.

The effect of an active third body on the potential well plays an important role on the fate of the molecule. Figure 1-2 displays the change of the potential well along the course of interaction of the intruding body "C" with AB. On point (a), along the path called reaction coordinate, C is too far away to have any effect on the depth and shape of the potential well of AB. In (b), where C is close enough to AB, the potential well becomes broader and its depth decreases. In (c) we are faced with a situation where both A and C have their maximum influence on B in the presence of each other. In this state, called the transition state, B is handing A over to C. The energy of this so-called activated complex $(ABC)^\ddagger$ along the reaction course is maximum and minimum along any other directions. In (d) the influence of A on B decreases and the influence of C on B increases. As a result the potential well of CA gradually becomes deeper and narrower along the reaction coordinate. In (e) the potential well represents the isolated CA.

The rate of the process depends on the energy difference between (ABC)‡ and AB, called activation energy, which in turn depends on the shape and spacing of the energy levels of the corresponding potential wells. The rate is slow if the potential well for AB is deep and narrow, or broad and shallow for (ABC)‡. The rate of recombination of B with AC to form AB (backward reaction, the process that B wants to take back A and kick C out), depends on the depths and widths of the AC and (ABC)‡ potential wells.

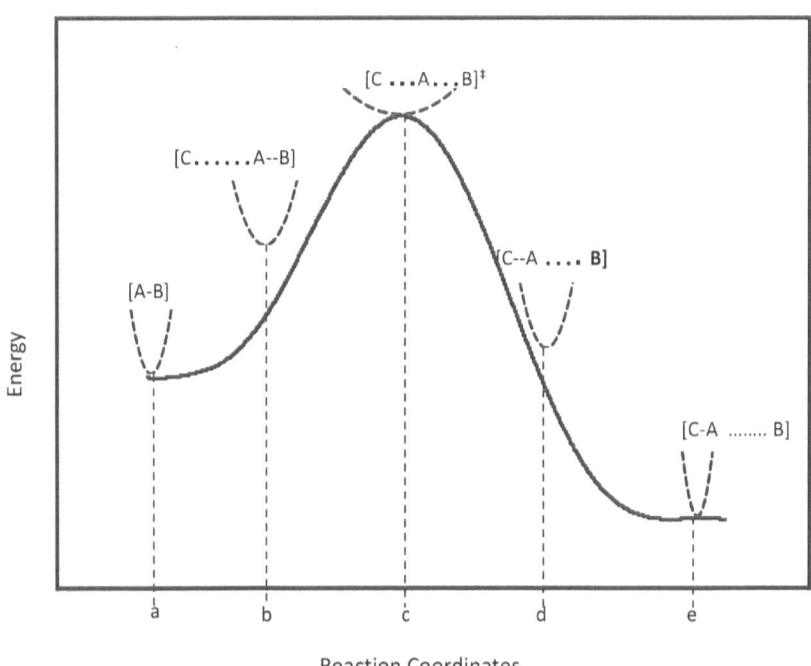

Figure 1-2 The influence of intruding body (C) on the shape and depth of the potential well describes the behaviour of the host body (AB) along the approaching course (reaction coordinates).

The ratio of forward rate reaction to the rate of backward reaction is called equilibrium constant which obviously is controlled by the depth and shape of the potential wells for AB and AC. Again the equilibrium constant, that is, the probability of AC formation is small if the potential well for AB is deep and narrow, or broad and shallow for AC.

The intruder in figure 1-2 could be a molecule. In this case the potential well for both molecules will change in a similar fashion (substitution reactions). In another category of chemical reactions, an atom or a molecule is added to another molecule forming a new molecule (addition reaction). The latter category of reactions occurs because of the nature of one atom (or atoms) in their parent molecule.

Many social structures such as family formation, and its stability under the influence of others and its rate of change, could be explained in terms of the above discussed reaction coordinates. If the intruder and a member of the family are in the right mood for interaction and the intruder approaches along the correct course then the family potential well will change in a manner similar to the manner presented in figure 1-2.

In the above classical interpretation of interaction of a third party with a member of a family, the third party not only should approach from a suitable direction, but should have enough energy to overcome the resistance. Yet at interstellar space at very low temperature, where the kinetic energy is very low and the species do not have enough kinetic energy to overcome the potential energy barrier, reactions are happening by means of a mechanism called tunneling. In the case of human behavior, it is called love at first sight.

In an open society where there exists a high probability of interactions between people, the shape and depth of the behavior well will change as the results of effective interactions. In this case, unlike in a closed society, the "Fate" that is the shape and depth of the potential behavior will change accordingly. Now Fate becomes a function of the will to interact and Chance. In analogy, it is expected that the rate of change to depend on the behavior well of the family and the behavior well of the new comer. The stability of the newly formed family and the tendency to return to the parent family depends on both behavior wells.

There exists another important class of potential wells called double minimum potential wells, which can be either symmetrical or unsymmetrical. Hydrogen bonded complexes are represented by double potential wells and proton transfer through the barrier separating the two wells are among the most fundamental chemical reactions. From the many interesting proton transfer reactions in a hydrogen bonded system that influence our lives, we will mention just one reaction; the double proton transfer in canonical base pair of DNA, in ground and excited electronic states. The proton transfer through the barrier is believed to be the cause of genetic mutation in the ground electronic state, and cause of radiation induced mutation, or carcinogenesis. Proton tunneling between two wells could be as fast as 10^9 sec^{-1}.

It is proposed that double behavior well could be attributed to dual personality, and transition from one personality to the other could be as fast as proton transfer between the two wells.

The potential wells outlined above describe the behavior and properties of molecules in their ground electronic state. Every molecule has many electronically excited states, and each state has its own characteristic potential well. Therefore, the chemical and physical properties of the molecules will be different in these electronic states. This means that a reaction that is not possible in one electronic state might be possible at another electronic state. Again, as a specific example consider Hydrogen molecules. The excited electronic states where the spins of the two electrons are anti parallel, as in the ground electronic state, are called singlet states, and the excited electronic states with parallel spins are called triplet states. Similar to transition between energy levels within a potential well, transition from an energy level in one electronic potential well to an energy level in any other electronic potential well is allowed if the energy of light packets is equal to the energy difference between two levels, and such a transition is allowed (determined by the properties of the two potential wells involved). The transition between a singlet state and triplet state is prohibited; however, the system could go from an excited singlet electronic state to a triplet state by a radiation-less mechanism, and then decay to the lower singlet ground electronic state by emitting light with a larger wavelength than that of the wavelength used for excitation (phosphorescence). The time scale for excitation from the singlet ground electronic state to a singlet exited electronic state, or relaxation to ground electronic state (florescence), is within 10^{-4} to 10^{-12} sec, and the life time for decay from a triplet state to singlet state is longer. Such a process is correlated with revelation with lasting effect.

Since the universe is made of matter, and these excited electronic states for each component of the universe exist within the matter, then the excited states of the universe exist within the universe itself and not around it. Recall that the shape and depth of the potential well for each component of the universe in these excited states is different from one another; therefore, the physico-chemical properties of these universes are different. In this description of seven heavens the transition from one universe to the other with the same multiplicity takes about 10^{-4} to 10^{-12} sec. The excited system will flash back to the ground state as fast as being excited, emitting light with frequency equal to that of excitation. However, if forbidden transition from one multiplicity in the excited universe to another multiplicity in the excited universe occurs, then the relaxation to the ground universe will be accompanied by illumination with much longer duration.

The potential energy well introduced above does not change with time, that is, the molecule is not under the influence of time dependent external potential energy. In cases where there exists an external potential energy as soon as the external potential energy is turned off, the potential energy well returns to its stationary state. But this not the case for human behavior, the behavioral well evolves with time. Therefore, in applying chemical laws to human behavior the time dependency of behavioral well should be taken into consideration.

Since computational chemistry has opened up a new frontier and has provided a tool for the chemist to predict the properties of known and unknown compounds and their molecular properties at ground electronic states and electronically excited states, then behavior computation may open up a new frontier for the prediction of behavior. As computational chemistry is based on some basis functions, behavior computation could be based on some basis behavioral functions to be optimized for each case.

Chapter Two

Microscopic Description of Matter

The concept of potential well and its role in determining the physico-chemical behavior of matter was introduced qualitatively in Chapter One. The mathematical description of potential well for real systems, and the extraction of physico – chemical properties of matter from the corresponding potential well, with the exception of very few simple systems, is very complicated and beyond the scope of this journey.

In order to become familiar with the mathematical extraction of the physico-chemical properties of matter from the potential well, here we introduce a simple one dimensional potential well which describes the behavior of a particle with mass "m" in an imaginary one dimensional potential space with the length of "L". It is assumed that the potential energy is constant within the boundaries of the well, and infinite outside of the well. With this restriction the probability of finding the particle outside of the boundaries is zero.

To describe the physical behavior of the system of interest in any given environment a mathematical function is needed, the function usually is represented by a letter of the Greek alphabet such as ψ, ϕ, χ , etc. In general such a function, real or imaginary, is a function of coordinates and time. If the function is real the probability of finding the particle at any point in space of interest is given by the squared value of this function at that point. If the function is imaginary, the probability of finding the particle at any point in space of interest is given by the value of the product of the complex conjugate of the function and the function at that point. Since the function is related to the probability then the function and its first and second derivatives should necessarily be single valued, and continues in the region of interest. In addition the sum of probabilities of the particle in the restricted region should be equal to 1. Such a function which satisfies the above three conditions is called a well-behaved function.

In addition to the function that describes the behavior of a particle in a given environment, for any physical property there exist an operator represented by a bold face letter such as P or $\hat{P}$. The physical property of interest can be obtained by operating the corresponding operator on the function describing the behavior of the particle. For example, for the energy of this particle in the above restricted space we notice that the potential energy within this space is constant and outside of this region is infinity. The particle is not under the influence of any external filed, so the energy of interest is the kinetic energy. The instructions for obtaining the operator for energy, or any other physical property, are given in supplementary section of this chapter. The operator for the kinetic energy in one dimensional space is,

$$\hat{H} = -\frac{\hbar^2 d^2}{2m dx^2}$$

Where $\hat{H}$ represents the operator for energy and it is called the Hamiltonian and $\hbar = h/2\pi$ is a constant and is called Plank's constant.

In order to extract energy of the system from function ψ_x the energy operator, that is $\hat{H}$ should act on the function ψ_x, that is;

$$\hat{H}\psi_x = E\psi_x$$

The above equation is an Eigen-function, Eigen-value equation known as Schrödinger or wave equation. It simply means that ψ_x is such a "well behaved function" that when operated on by $\hat{H}$, will give a scalar (energy) multiplied by itself.

For this simple case, there are few simple functions that have the above property, that is when one takes the second derivative of the function, one obtains the function back multiplied by a number, in this case energy. Sinuous function is one of them, so we choose the function $\psi_x = A\sin\alpha x$ as the function describing the behavior of the particle in one dimensional potential space.

Since this book is intended for those with little knowledge of mathematics, the mathematical calculations will be presented in the supplementary section of this chapter, and here we will discuss the results of the above mentioned operations in a qualitative manner for this case.

The requirement of being a well behaved function leads to the following results,

$$E_n = \frac{n^2\pi^2\hbar^2}{2mL^2} = \frac{n^2h^2}{8mL^2}$$

and

$$\psi_{n,x} = \sqrt{\frac{2}{L}}\sin\frac{n\pi}{L}x$$

Where "n" called quantum number is an integer, $n = 1, 2, 3, \ldots$. This means that at microscopic level, energy is not continuous, but it has discrete values, or energy is quantized. These quantized energies are determined by the mass of the particle and the dimension of the potential space that the particle is restricted to. It should be emphasized that function ψ does not have any physical meaning, only the square of this function, if it is real, like in this case, or if imaginary, product of the complex conjugate of the function and the function has physical meaning.

For $n=1$, the state function is given by $\psi_{1x} = \sqrt{\frac{2}{L}} \sin \frac{\pi}{L} x$, and its energy given by $E_1 = \frac{h^2}{8mL^2}$. This simply means that if the energy of the particle is E_1, then the probability of the particle being at $x = 0$, and $x = L$ (the borders) is zero and in the middle is maximum. For $n = 2$, the particle is described by $\psi_{2x} = \sqrt{\frac{2}{L}} \sin \frac{2\pi}{L} x$ and its energy $E_2 = \frac{h^2}{2mL^2}$, referred to the first excited state. In addition to the borders the probability of the particle being in the middle is zero, and probability of the particle being at $\frac{1}{4}L$ and $\frac{3}{4}L$ is maximum.

Now, that the function describing our particle in a restricted one dimensional potential space is available, we can calculate any physical property of the system by operating the associated operator of the physical property on the function, provided that the Eigen-function- Eigen-value relation holds. If not, then we can calculate the average value of that property at any point. For example The Eigen-value-Eigen Function relation does not hold for position, momentum and square of position, but we can calculate their average values. From the average values for position, momentum and square of position and exact value of square of momentum, we can calculate the variance and standard deviation in position and momentum to demonstrate an important property of the microscopic systems, i.e., Heisenberg uncertainty principals. Again we will present the mathematical derivation in the supplementary section of this chapter for the interested reader. If one multiplies the standard deviation in momentum σ_P with the standard deviation on position σ_x, one obtains the following relation;

$$\sigma_x \times \sigma_P \approx \hbar/2,$$ where, "$\approx$" means in the order and $\hbar$ is Plank's constant $= \dfrac{h}{2\pi} = 1.05457126(47) \times 10^{-34}$ J.S.

This relation simply means that if one wants to determine the exact position of the particle, i.e. $\sigma_x = 0$, then one loses the information about the momentum, i.e. σ_P becomes infinite, and vice versa.

Although the uncertainty principal is meaningless in the macroscopic domain, in lieu of our general discussion in Chapter One, this principal could be applied to human behavior. As a general example consider a group of people who believe in a set of principles. If we show the diversity of "opinion" on the principals by $\Delta\, OP$ and the scope of these principals by $\Delta\, SP$ then:

$\Delta\, OP \times \Delta\, SP = Constant$

Which means for minimum diversity in opinion the scope of the principals should be enlarged to cover the opinion of everyone in the group, then as $\Delta\, OP \to \infty$ the scope goes to zero which simply means they do not agree on anything. This principal could be applied to any two conjugated behavioral patterns.

Now, let us calculate the energy difference between energy level E_2 and E_1, that is;

$$\Delta E = E_2 - E_1 = \frac{h^2}{2mL^2} - \frac{h^2}{8mL^2} = \frac{3h^2}{8mL^2}$$

If one photon with energy exactly equal to ΔE, interacts with this particle satisfying certain rules called selection rules, then this photon will be absorbed by the particle at energy level E_1, and the particle will be excited to energy level E_2. The energy of this photon is given by $E = h\nu = \frac{hc}{\lambda}$, where ν is the frequency and λ is the wavelength.

Now imagine that we have a source of photons with different frequencies (wavelengths), including the one that its energy matches the energy difference between the energy levels E_2 and E_1, that is, $\frac{3h^2}{8mL^2}$ then one photon will be absorbed by this particle, and if there are "n" none reacting particles, then n photons with energy $\frac{3h^2}{8mL^2}$ will be absorbed by these particles. If a detector is set in such a position that measures the frequency or the wavelengths of the photons passing through, then the intensity of the photons with frequency ν will be reduced equal to the number of particles. Therefore, one can use the relation $h\nu = \frac{hc}{\lambda} = \frac{3h^2}{8mL^2}$ to calculate the length of the potential well and thus the energy levels. The particles in the excited state are not in a stable state (called first excited state), so it will lose energy in the form of photons, and decay back to energy level E_1 (called the ground state). The photons of n body system will be emitted in different directions and so if a detector is placed in an angle to the excitation light (source of photons), by measuring the frequency of the emitted photons one can obtain the same information as obtained in the former experiment. The former is called absorption spectroscopy and the latter emission spectroscopy.

The Two-Dimensional Problem

So far we have been concerned with the problem of a particle in imaginary one dimensional potential space. Now we will consider the case of a particle in an imaginary two dimensional potential space (X,Y). Within the boundaries of L_x and L_y, the potential is constant and outside of the boundaries the potential is infinity. Since potential energy is constant in this case, we are only concerned with kinetic energy. The operator for the kinetic energy in two-dimensional space is

$$\hat{H} = \frac{-h^2}{2m}\left(\frac{d^2}{dx^2} + \frac{d^2}{dy^2}\right)$$

Whenever the operator can be written in terms of independent operators, the function can be written in terms of products of independent functions. The technique is known as the separation of variables, and by using this technique, the two dimensional problem reduces to two one dimensional problems, which we already are familiar with.

$$\psi_{(x,y)} = f_x \times g_y$$

Where by imposing the boundary conditions and the requirement of well behaved functions, we have,

$$f_x = \sqrt{\frac{2}{L_x}}\sin\frac{n_x\pi}{L_x}x, \quad E_{n(x)} = \frac{n_x^2 h^2}{8m L_x^2} \quad \text{and}$$

$$g_y = \sqrt{\frac{2}{L_y}}\sin\frac{n_y\pi}{L_y}y, \quad E_{n(y)} = \frac{n_x^2 h^2}{8m L_x^2}$$

Notice that now there are two quantum numbers (n_x) and (n_y). Therefore,

$$\Psi_{(x,y)} = \frac{2}{\sqrt{L_x L_y}} \sin\frac{n_x \pi}{L_x} x \, \text{Sin}\frac{n_y \pi}{L_y} y \quad \text{and}$$

$$E_{n_{(x)} n_{(y)}} = \frac{h^2}{8m}\left(\frac{n_x^2}{L_x^2} + \frac{n_y^2}{L_y^2}\right)$$

For $n_x = 1$ and $n_y = 1$

$$E_{1,1} = \frac{h^2}{8m}\left(\frac{1}{L_x^2} + \frac{1}{L_y^2}\right) \quad \text{and} \quad \Psi_{(x,y)} = \frac{2}{\sqrt{L_x L_y}} \sin\frac{\pi}{L_x} x \, \text{Sin}\frac{\pi}{L_y} y$$

For $n_x = 1$ and $n_y = 2$

$$E_{1,2} = \frac{h^2}{8m}\left(\frac{1}{L_x^2} + \frac{4}{L_y^2}\right) \quad \text{and} \quad \Psi_{(x,y)} = \frac{2}{\sqrt{L_x L_y}} \text{Sin}\frac{\pi}{L_x} x \, \text{Sin}\frac{2\pi}{L_y} y$$

$$E_{2,1} = \frac{h^2}{8m}\left(\frac{4}{L_x^2} + \frac{1}{L_y^2}\right) \quad \text{and} \quad \Psi_{(x,y)} = \frac{2}{\sqrt{L_x L_y}} \text{Sin}\frac{2\pi}{L_x} x \, \text{Sin}\frac{\pi}{L_y} y$$

Which means the energy of the two excited states described by two different functions are different. If $L_x = L_y$ then the energy for the two different functions;

$$\Psi_{(x,y)} = \frac{2}{L} \text{Sin}\frac{\pi}{L} x \, \text{Sin}\frac{2\pi}{L} y \quad \text{and}$$

$$\psi_{(x,y)} = \frac{2}{L} Sin\frac{2\pi}{L} x \; Sin\frac{\pi}{L} y$$

Will be,

$$E_{1,2} = E_{2,1} = \frac{5h^2}{8mL^2}$$

If two distinct functions correspond to the same energy, then the energy states are degenerate. If one applies an external potential in one direction then $L_x \neq L_y$ and the degeneracy is removed.

Finite Potential Well

In chapter one, we referred to an important class of potential wells called double minimum potential wells, where protons could penetrate through potential barriers, a phenomenon that has no counterpart in classical mechanics. In order to understand the basic concept of tunneling we will discuss the case of finite potential well. To simplify the discussion further we will focus on a potential well with infinite potential at $x = 0$ and finite potential V_0 at $x = L$. Between $x = 0$ to $x = L$ the problem is similar to the case of particle in one dimensional space with infinite potential, therefore;

$$\psi_x = A Sin(\alpha x) \qquad \alpha = \frac{\sqrt{2mE}}{h}$$

In the region from $x = L$ to $x = \infty$

$$\varphi_x = BSin(\beta x) \qquad \beta = \frac{\sqrt{2m(E-V_0)}}{h}$$

For the case of $E < V_0$, that is the energy of the particle is less than E, the boundary condition requires that ψ_L φ_L. There is no analytical solution for this equation and it should be solved numerically. The numerical solution for this equation shows that ψ_n, is not zero at region $> L$ and $E < V_0$. This phenomenon is called tunneling. A new approach for calculation of tunneling through a barrier with finite widths could be found in professor Gilfoyle lectures from University of Richmond, Virginia[4],

Application of quantum chemistry to the ground and excited electronic states of biological molecules has opened up a new field referred to as Quantum Biology. Proton tunneling through potential barriers of hydrogen bonds between two complementary bases in DNA could lead to a change in one of the pair bases. The tunneling probability not only depends on base pair, but also on the electrostatic of environment, and other factors.

Graphical Presentation of State and Probability Functions

The graphical representation of $\psi_{1,x} = \sqrt{\frac{2}{L}} \sin\frac{\pi}{L}x$ is shown in figure 2-1. In this graphical representation, coordinate Ψ, shows only the value of Ψ at any point between $0 \geq x \leq L$. For $\psi_{1,x} = \sqrt{\frac{2}{L}} \sin\frac{\pi}{L}x$

---

[4] A New Teaching Approach to Quantum Mechanical Tunneling, G.P. Gilfoyle. Physics Department, University of Richmond, Virginia, 23173. GILFOYLE@URVAX.URICH.EDU

the function has a positive value at any point. For $\psi_{2,x} = \sqrt{\dfrac{2}{L}}\,\sin\dfrac{2\pi}{L}x$, the value of Ψ at any point between $0 \geq x \leq L/2$ is positive and at any point between $L/2 \geq x \leq L$ is negative. It should be emphasized that Ψ and its sign do not have any physical meaning. However, for the cases where the potential is not constant, for example for variable potential energy in this example or in multi-electron systems the problem cannot be solved exactly, and approximation methods are used for solving the Schrödinger equation. In approximation methods a linear combination of known state functions, called base functions is chosen as the solution. The coefficients in this linear combination of base functions are optimized in such a way that energy for that set of functions becomes a minimum. Organic chemists prefer to avoid the theoretical calculations; instead they use graphical representation of the linear combinations of state functions to acquire an approximate representation of the state function for their system of interest.

For the state function $\psi_{1,x}$ the probability of particle being at $0, \dfrac{L}{4}, \dfrac{L}{2}, \dfrac{3L}{4}$ and L and for the state function $\psi_{2,x}$ are represented by bars on the bottom-right and top-right of Figure 2-1.

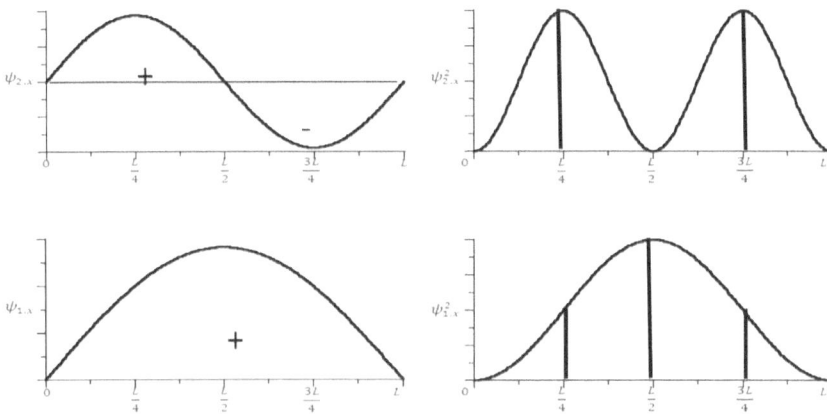

Figure 2-1 Graphical representation of $\psi_{1,x} = \sqrt{\dfrac{2}{L}} \sin \dfrac{\pi}{L} x$ and For $\psi_{2,x} = \sqrt{\dfrac{2}{L}} \sin \dfrac{2\pi}{L} x$ (bottom-left and top-left respectively), + The plus and negative signs have no physical meaning and only refer to the sign of the state function at the specified region. The probability functions, $\psi_{1,x}^2$ and $\psi_{2,x}^2$ are shown in the bottom-right and bottom-left.

Chapter Two Supplementary

Instructions for Obtaining Quantum Mechanical Operators for Observable Physical Properties

For any physical property there exist an operator represented by a bold face letter such as **P** or $\hat{P}$. In order to obtain this operator, we will use the following simple rule:

1- Write the classical function that classically describes that property.
2- Leave any function of the coordinates unchanged.
3- Replace the momentum by, $-i\hbar\dfrac{\partial}{\partial q}$ where q stands for general coordinates, $i = \sqrt{-1}$ and $\hbar = h/2\pi$ is Plank's constant.

As an example let us consider the operator for total energy, called Hamiltonian, shown by symbol **H** or $\hat{H}$. Here we will be concerned only with kinetic energy and potential energy, that is, $T = \tfrac{1}{2}mv^2 + V_x$. Kinetic energy could be expressed in terms of momentum; therefore, the classical expression for kinetic energy in one dimension is:

Since $P = mv$ Then,

$$T = \frac{(mv)^2}{2m} = \frac{P^2}{2m}, \text{ and}$$

$$\hat{T} = \frac{1}{2m}\left(-i\hbar\frac{\partial}{\partial q}\right) \times \left(-i\hbar\frac{\partial}{\partial q}\right) = \frac{-\hbar^2 \partial^2}{2m\,\partial q^2} \quad \text{S2-1}$$

Since potential energy depends on coordinates and not momentum, the expression for potential function is left unchanged. The quantum expcs,

$$\hat{H} = \hat{H} + V_{(q)} \quad \text{S2-2}$$

Case of a Particle in One Dimensional Potential Space

In this case $\hat{H}\psi_x = E\psi_x$ becomes,

$$\frac{-\hbar^2}{2m}\frac{d^2\psi_x}{dx^2} + V\psi_x = E\psi_x \quad \text{S2-3}$$

Since V is constant, it can be set to zero, which simply means we are measuring the energy of the system with respect to this constant. In this case equation S2-3 becomes,

$$\frac{d^2\psi_x}{dx^2} = \frac{-2mE}{\hbar^2}\psi_x$$

The quantity $\dfrac{2mE}{h^2}$ is a constant, for simplicity we will replace it by α^2, thus

$$\frac{d^2\psi_x}{dx^2} = -\alpha^2 \psi_x \qquad \text{S2-4}$$

We are seeking a well behaved function when operated by operator "$\dfrac{d^2}{dx^2}$" that is, taking the second derivative of the function with respect to x will produce the function itself multiplied by a scalar. We can easily identify some simple well behaved functions for this case, for example, $A\sin(\alpha x)$, $B\cos(\alpha x)$ or a combination of the two, $Ce^{-i\alpha x}$, $De^{i\alpha x}$, or a combination of the last two exponential functions, which by the Euler equation is equivalent to the combination of *sine* and *cosine* functions. The constants A, B, C, and D could be determined from the properties of the function. To continue, we choose $\psi_{(x)} = A\sin(\alpha x)$. The first derivative of $\psi_{(x)}$ with respect to x, gives $\alpha A\cos(\alpha x)$ and the first derivative of this function or the second derivative of the original function with respect to x gives $-\alpha^2 A\sin(\alpha x)$.

For outside of the boundaries since $V = \infty$ then the probability of finding the particle in this region is zero which simply means that $\psi_{(x)} = 0$ in this region. Since $\psi_{(x)}$ is a continuous function of x and it is singled value, then $\psi_{(x)}$ should be zero at the two boundaries, that is at $x = 0$ and at $x = L$,

$$\psi_{(0)} = A\sin(0) = 0 \text{ and,}$$

$$\psi_{(a)} = A\sin(aL) = 0$$

The first equation is true by the properties of the sine function. The second equation will hold only if,

$$\alpha L = n\pi \text{ or } \alpha^2 L^2 = n^2 \pi^2 \quad n = 1, 2, \ldots.$$

Since $\alpha^2 = \dfrac{2mE}{h^2}$ then,

$$E_n = \frac{n^2 h^2 \pi^2}{2m} \qquad \text{S2-4}$$

Notice that energy depends on n, since n is an integer, then energy has discrete values.

Other Physical Properties of the System

The position of the particle in the potential space and its momentum are two properties of interest. The operator for position in one-dimensional space is $\hat{x} = x$ and for momentum is, $\hat{P}_x = -i\hbar \dfrac{\partial}{\partial x}$. In the first case;

$$\hat{x}\psi_x = \sqrt{\frac{2}{L}} x \sin \frac{n\pi}{L} x$$

And for the second case;

$$\hat{P}_x \psi_x = \sqrt{\frac{2}{L}} \left(-i\hbar \frac{\partial}{\partial x}\right) \sin \frac{n\pi}{L} x = -i\hbar \frac{n\pi}{L} \sqrt{\frac{2}{L}} \cos \frac{n\pi}{L} x$$

In the above two cases Eigen-function, Eigen-value relations do not hold. Therefore, in accordance with the postulates of quantum mechanics (which we have not explicitly mentioned them neither in chapter two, nor in this supplementary section), in cases where Eigen-function, Eigen-value do not hold, the average value of the physical property (in these cases position and momentum and the square value of the position x^2) could be calculated from the following relations:

$$\langle x \rangle = \frac{\int_0^a \psi_x^* \hat{x} \psi_x dx}{\int_0^a \psi_x^* \psi_x dx} \qquad \text{S2-5}$$

$$\langle x^2 \rangle = \frac{\int_0^a \psi_x^* \hat{x}^2 \psi_x dx}{\int_0^a \psi_x^* \psi_x dx} \qquad \text{S2-6}$$

and

$$\langle P_x \rangle = \frac{\int_0^a \psi_x^* \hat{P}_x \psi_x dx}{\int_0^a \psi_x^* \psi_x dx} \qquad \text{S2-7}$$

For the square of momentum P^2 Eigenfunction – Eigenvalue relations hold, so

$$P^2 = (n\pi h / L)^2 \qquad \text{S2-8}$$

Equations S2-5, S2-6 and S2-7 can be solved by using the following relations:

$$\int x \sin^2 \alpha x \, dx = \frac{x^2}{4} - \frac{x \sin(2\alpha x)}{4\alpha} - \frac{x \cos(2\alpha x)}{4\alpha^2}$$

$$\int x^2 \sin^2 \alpha x \, dx = \frac{x^3}{6} - \left(\frac{x^2}{4\alpha} - \frac{1}{8\alpha^3}\right) - \frac{x\cos(2\alpha x)}{4\alpha^2}$$

Then,

$\langle x \rangle = \dfrac{L}{2}$, an obvious result

$$\langle x^2 \rangle = L^2 \left[\frac{1}{3} - \frac{1}{2(n\pi)^2}\right]$$

And

$$\langle P_x \rangle = 0$$

Variance is defined as the square deviation from the square of the mean, so

$$\sigma_x^2 = \langle x^2 \rangle - \langle x \rangle^2 = L^2 \left[\frac{1}{12} - \frac{1}{2(n\pi)^2}\right] \qquad \text{S2-9}$$

$$\sigma_P^2 = \langle P^2 \rangle - \langle P \rangle^2 = (n\pi h/L)^2 \qquad \text{S2-10}$$

Standard deviation is the square root of the variance, and the product of standard deviation in momentum and position is of interest.

$$\sigma_x \sigma_p = \frac{h}{2}\left[\frac{(n\pi)^2}{3} - 2\right]^{1/2} \quad \text{for } n = 1$$

$$\sigma_x \sigma_p = 1.13572 \left(\frac{h}{2}\right) \qquad \text{S2-11}$$

Chapter Three

Macroscopic Description of Matter Behavior

Classical thermodynamics

In chapter two, we viewed the molecular behavior of matter through the window of a simple one-dimensional potential well. In this chapter, unlike in the previous two chapters, we will study the behavior of matter as an ensemble of molecules (atoms). Such a system with thermal, mechanical, electromagnetic properties is very complicated; therefore, it is simpler to restrict the study to simple systems. By simple system we mean those systems that are homogenous, chemically inert, and electrical and magnetic fields have no effect on them. This restriction will not limit the generality of the theory that will be discussed in this chapter; the interested reader is referred to the excellent book by Herbert B. Callen [5]. For other case the restriction for each case could be removed simply by adding the term related to that property available from the theory describing

---

[5] Thermodynamics, an introduction to the physical theories of equilibrium thermostatic and irreversible thermodynamics by Herbert B. Callen, John Willey and Sons, Inc. 1960. The theoretical formulism presented in this chapter and supplementary section is a simplified version of Callen's formalism.

that property. This simple system is described by three independent parameters. The number of molecules of each component (or number of moles) and volume are two obvious parameters. For the other parameter we choose the internal energy.

Internal energy

A macroscopic system is an ensemble of many electrons and nucleons with a complicated energy of interaction between them. We might not be able to measure this interaction energy, but this energy for any given state of the system is constant, and obeys the principal of conservation of energy. It should be emphasized that the energy difference between two states of a system has physical meaning, and not its absolute value. Usually one state of a system is chosen as the reference state, and energy of the system in any other state with respect to the reference state is called the internal energy of the system, which is represented by symbol E or U.

Extensive parameters

The above three parameters have a very important property in common. Suppose that a composite system is made of two systems exactly alike. Then the volume of the composite system will be exactly two times of the volume for each system. The same is true for the number of molecules and internal energy of the system.

Any parameter that has the above property is called extensive parameter. These types of parameters play an important role in thermodynamics.

Thermodynamic Equilibrium

In this Journey we will limit our discussion of thermodynamics to simple systems, but there exist another more serious limitation, that limits the subject under discussion to those states of the systems that are called equilibrium states. Here we will not get involved in non-equilibrium thermodynamics. As an example of a non- equilibrium state consider the state of a stationary liquid. It is much simpler to describe this state than the state of a liquid system that flows. In order to express the equilibrium state of the system we will state the following.

The equilibrium state of a simple macroscopic system is completely described by three extensive parameters that are the number of molecules (moles), the volume, and the internal energy. When the restriction of a simple system is being removed, the number of parameters will increase accordingly.

Walls and Constraints

To describe a thermodynamic system, it is necessary that the type of walls that separate the system from its surroundings be specified. A thermodynamic process occurs because the type of wall changes in that process. For example, consider a gas system composed of two simple systems in a solid cylindrical tube, separated by a piston. If the piston is fixed, then the volume of each system is fixed, but if the constraint is removed, then the piston will move until the system reaches its new equilibrium state. A wall that permits the volume to change is called a permeable wall with respect to volume, and if the wall is fixed then it is called an impermeable wall with respect to volume. In general a wall that permits an extensive parameter to change is called permeable with respect to that parameter and if not, it is called an impermeable wall. A wall that restricts the flow of energy in the form of heat to the system is called an adiabatic wall, and if it does not restrict the flow of energy in the form of heat it is called a diathermal wall.

If we surround a system with impermeable walls with respect to the number of molecules and adiabatic walls, then the number of molecules in the system of interest will not change and energy cannot be transferred from the system to the surrounding by means of heat. The only way that energy of this system could change is by work energy. If work is done on the system then the energy of the system will increase and if work is being done by the system then the energy of the system will decrease. In this case the change of energy from state A to B will be

$$\Delta U = U_B - U_A = W \qquad \text{3-1a}$$

By replacing the adiabatic wall with a diathermal wall and keeping the volume constant, we can transfer the system from state A to state B, only by letting energy in the form of heat to flow from the surroundings, to the system. In this case,

$$\Delta U = U_B - U_A = Q \qquad \text{3-1 b}$$

Notice that in the above two paths, the energy difference between state A and state B is independent of the path, but work and heat depend on the path. If we generalize this concept, then,

$$\Delta U = U_B - U_A = W + Q \qquad \text{3-2}$$

What equation 3-2 implies is that there are infinite pathways between the states A and B. For infinitesimal change;

$$dU = đW + đQ \qquad \text{3-3}$$

Where đ shows that infinitesimal change for work and heat is path dependent. This is the first law of thermodynamics.

In Chapter 1 we claimed that there exists a correlation between human behavior and molecular behavior. Thermodynamics describes the macroscopic behavior of an ensemble of molecules and thus in a qualitative manner it is capable of describing the behavior of an ensemble of people. In this case walls and constraints in any society are defined by the moral beliefs, principals, and laws imposed over that society. As long as the walls and restraints are in place, then the society is in a state of equilibrium, but as soon as one of the constraints is removed, the society will change from its original state of equilibrium to a new equilibrium state.

The Entropy Postulates

Although the mathematical formulation in chapter two was based on some postulates, in order to keep the discussion as simple as possible we intentionally did not mention them. The thermodynamic postulates are much easier to understand, and therefore we will explicitly present them here. The postulates are accepted without proof. Laws are derived from postulates and these laws can explain the physical observations and enable one to make predictions. These postulates like any other postulates in science, and for that matter in any other disciplines, cannot be proven. If one can prove any of these postulates then he has to use a law that has been derived from other postulates.

Postulate I

The equilibrium state of any composite system is described by a state function of extensive parameters called entropy "S". At any equilibrium state the extensive parameters will have the values that maximize the entropy in that equilibrium state.

It should be emphasized that these postulates hold only for equilibrium states. This postulate is sometimes referred to as the maximum entropy principal. The relation between entropy and extensive parameters is called the fundamental relation of thermodynamics.

Postulate II

This postulate has two parts.

A) Entropy of a composite system is additive over the subsystem entropies.
B) Entropy is a continuous and differentiable function and monotonically increases with energy.

From this postulate we can deduce some mathematical conclusions. Part A could be expressed mathematically by

$$S = \sum_{\alpha} S^{\alpha} \qquad 3\text{-}4$$

Where S^{α} represents the entropy of subsystem α.

We will present the mathematical development of classical thermodynamics in Chapter Three's Supplement and will continue our discussion in a qualitative manner.

According to postulate **I,** entropy is a function of extensive parameters; this means the entropy is an extensive parameter. The monotonic property means that the partial derivative of entropy with respect to energy is a positive quantity. In supplementary section it will be shown that the reverse of this partial derivative matches the definition of temperature, then this postulate implies that temperature (absolute) is a positive quantity.

The continuity, differentiability, and monotonic properties allows that the function to be reversed with respect to energy. Then, energy will be a continuous and differentiable function of S, V, and n. That is, equation:

$$S = S(U,V,n_1,n_2,...),\qquad 3\text{-}5$$

could be solved for U

$$U = U(S,V,n_1,n_2,...)\qquad 3\text{-}6$$

Equations 3-5 and 3-6 both represent different forms of fundamental relation and both, if known, contain all the thermodynamic information about the system.

Postulate III

The entropy of any system vanishes in the state for which

$$\left(\frac{\partial U}{\partial S}\right)_V = 0 \qquad 3\text{-}7$$

That is at absolute zero, entropy is zero.

Intensive parameters

The fundamental relation in the form of 3-5 or 3-6, if known, contains all thermodynamic information about the system, but the fundamental relation cannot be deduced from thermodynamics. Thermodynamics give information about the change from one equilibrium state to another. Therefore, the differential form of the fundamental relation plays the leading role in thermodynamics. If we write the fundamental relation for one component system in the form of energy

$$U = U(S,V,n)$$

Then the first derivative is;

$$dU = TdS - PdV + \mu dn \qquad \text{3-8}$$

Where, the partial derivatives of energy with respect to entropy, volume and mole number are defined as temperature, pressure, and chemical potential respectively. For detail see the Chapter Three supplement.

Reversible and irreversible processes

We have emphasized at the beginning of this chapter that thermodynamics is the study of equilibrium states. Nonetheless, the process that takes the system from one equilibrium state to another is of great significance. Again we like to emphasis that thermodynamics is not capable of providing any information about the dynamics of the path between two equilibrium states.

Let's assume that a closed system (system 1) is equilibrium state "A". By removing one constraint on this system, the system spontaneously will reach to a new equilibrium state "B", because entropy in state "B" is greater than that of state "A", if one wishes to restore this system to its original state "A" one cannot accomplish this wish simply by manipulating constraints within the system, because entropy in state "A" is less than entropy in state "B". However, there exists a way that constraints of state "A", could be imposed on the system 1. Consider another system (system 2) in equilibrium state "C". By removing a constraint from this system, this system will reach a new state of equilibrium "D". Let's assume that the entropy change for the process in system 2 is larger than the entropy change from state "B" to state "A" in system 1. Now if we couple these two systems, then we have a composite system in which the entropy is reducing in one section and increasing in the other section. Under these conditions, system 1 will be restored to its original state, because the entropy of the composite system has increased. Again one cannot restore this composite system to its original conditions; that is state "A" to "B" in system 1 and state "D" to state "C" in system 2, because entropy cannot decrease unless we couple this composite system with another system in which entropy increases more than entropy decrease in the composite system. In general for a real process;

$$\sum_\alpha S^\alpha > 0 \qquad \text{3-9a}$$

To continue our discussion of societies in terms of thermodynamics if a constraint is removed, then the society will chose a new state of equilibrium. It is impossible to take back this society to its original state by imposing the original constraints on them because it requires a reduction of entropy. However, it is possible to impose constraints on the society, but it should be accompanied by the removal of other constraints in the society in such a way that the increase in entropy is more than the reduction of entropy for imposing the desired constraint.

There exists an ideal process called "reversible process". In this idealistic process, system 1 passes through infinite equilibrium states from state "A" to state "B" and so does system 2 from state "C" to state "D" in system 2. Thus, the decreases in entropy in system 1 are exactly equal to increases in entropy in system 2. That is,

$$\Delta S = \Delta S^{(1)} + \Delta S^{(2)} = 0 \qquad \text{3-9b}$$

Such a composite system could go in either direction. In the supplementary section of this chapter we will show that the efficiency of this reversible process is maximum in this idealistic process and as the process deviates from this ideal process, the efficiency decrease. Indeed there exists one pathway for which the efficiency is zero. This brings us to another conclusion.

There is a Beginning and there is an End.
There exist an infinite number of pathways between the beginning and the End.
No matter what route one chooses, the End will be the same.
Among the infinite number of pathways there is one pathway, called reversible pathway (Heavenly way) with maximum efficiency.
There is another pathway for which the efficiency is zero (Hell way).
All other pathways are combination of these two extreme cases.
The closer the chosen path is to the ideal path, the more rewarding the journey will be.

An Alternative Representation of Thermodynamic Fundamental Relations

Legendre Transformation

In both energy and entropy representations of fundamental relations, extensive parameters play the role of independent parameters, and intensive parameters are introduced as the first derivative. However, it is much easier to measure and control the intensive parameters than extensive parameters. It is possible by a mathematical transformation, known as the Legendre transformation, to transform the fundamental relation in such a way that temperature replaces entropy and pressure replaces volume. The mathematics of this transformation is discussed in the supplementary section of this chapter. Here, we will use the results of the Legendre transformation.

Replacing entropy with temperature in the fundamental relation leads to a new function called Helmholtz potential or Helmholtz free energy donated by F or A.

$$F = U - TS \qquad 3\text{-}10$$

The complete differential of F is,

$$dF = dU - SdT - TdS \qquad 3\text{-}11$$

By substituting for dU from equation 3-8 we obtain;

$$dF = SdT - PdV + \mu dn \qquad 3\text{-}12$$

At constant temperature and n, equation 3-12 becomes;

$$dF = dU - TdS$$

Since $dU = \text{\cancel{$dQ$}} + \text{\cancel{$dW$}}$

Then in an ideal reversible process at constant temperature and n,

$$\Delta F = W \qquad 3\text{-}13$$

Or

$$-\Delta F = -W = W'$$

W' stands for the work done by the system. It should be emphasized that by 'constant temperature' we mean that the temperature at the start and finish are the same, and it does not mean that the process follows a constant temperature path. Since F is a state function independent of path, then one can calculate the maximum ideal attainable work at constant temperature for a process.

Enthalpy

Another relation designated by "H", called Enthalpy is obtained by partial Legendre transformation of fundamental relation by replacing V with its conjugate intensive parameter "$-P$".

$$H = U + PV \qquad \text{3-14}$$

The differential form is;

$$dH = dU + PdV + VdP$$

Substituting for dU

$$dH = TdS + VdP + \mu dn \qquad \text{3-15}$$

At constant pressure and constant n equation 3-15 becomes;

$$dH = TdS$$

Gibbs free energy

Replacing both entropy and volume in fundamental relation with their corresponding intensive conjugates parameters in a Legendre transformation produce a function called Gibbs free energy designated by "G"

$$G = U + PV - TS \qquad \text{3-16}$$

The differential form is;

$$dG = dU + PdV + VdP - TdS - SdT$$

Substituting for dU

$$dG = -SdT + VdP + \mu dn \qquad 3\text{-}17$$

Since $dU = đQ + đW$, and we can write $đW$ as a sum of mechanical work, $đW = -P_{ex}dV$ and other types of work, W'. Then, in an ideal reversible process at constant temperature and pressure (again it should be emphasized that by constant temperature and pressure we mean the temperature and the pressure of the start and finish are the same, and it does not mean that the process follows a constant temperature and pressure path) we can write

$$\Delta G = W' \qquad 3\text{-}18$$

Since G is a state function independent of path, we can then calculate the maximum attainable work for a process for which the final temperature and pressure is the same as the initial state.

The relation between $\Delta U, \Delta H, \Delta F$ *and* ΔG *finds a qualitative application in our daily life.* ΔU *is correlated with the income "* $\Delta(inc)$*", a state function that depends on many parameters defining a person, such as, education, training, ambition, environment, etc. Like* ΔU*, which in our simple system varies with temperature and volume,* $\Delta(inc)$*, varies with change of any parameter that define the person. But for fixed parameters* inc *is constant.*

Now according to the postulate of maximum entropy some of this income should be dissipated as tax. Some part of the remaining income is used for living expenses. What is left, like Gibbs free energy is used for pleasure, etc.

Specific Heats and Other Derivatives

The second derivatives of fundamental equations play important role in thermodynamic formulization. These second derivatives, like first derivatives, can be measured by laboratory techniques.

The coefficient of thermal expansion is defined by,

$$\alpha \equiv \frac{1}{V}\left(\frac{\partial V}{\partial T}\right)_{P,n}$$

This coefficient is a measure of increase in the volume per unit increase in temperature at a constant pressure (and constant mole numbers).

The isothermal compressibility is defined by:

$$\beta \equiv -\frac{1}{V}\left(\frac{\partial V}{\partial P}\right)_{T,n}$$

This coefficient is a measure of decrease in volume per unit increase in pressure, at constant temperature (and constant mole numbers).

The specific heat at constant pressure is defined by:

$$C_P \equiv T\left(\frac{\partial S}{\partial T}\right)_P = \left(\frac{đQ}{dT}\right)_P$$

And the specific heat at constant volume is defined by:

$$C_V \equiv T\left(\frac{\partial S}{\partial T}\right)_V = \left(\frac{đQ}{dT}\right)_V$$

The Molecular Concept of Entropy

We are now in the position to introduce the concept of entropy at molecular level. For this purpose we reconsider the case of N body system in one dimensional potential space for N=4 particles. Furthermore, we assume that only three energy levels are accessible for each particle (case A), and the system somehow has acquired an energy equal to $E_t = 15\varepsilon_1$, where ε_1 is the energy of the first level described by state function φ_1 (Symbols E and ε refer to energy total energy and individual energy at molecular level and symbol U to energy at the macroscopic level). This means that two of the particles are described by state function φ_1, the third by state function φ_2 with energy $\varepsilon_2 = 4\varepsilon_1$ and the fourth by state function 3 with energy $3 = 9\varepsilon_1$. The state function for this 4 body system is $\psi_t = \varphi_1 \times \varphi_1 \times \varphi_2 \times \varphi_3$. There are 12 ways that these 4 particles could be distributed with equal probability between available energy levels as shown in the figure 2-1.

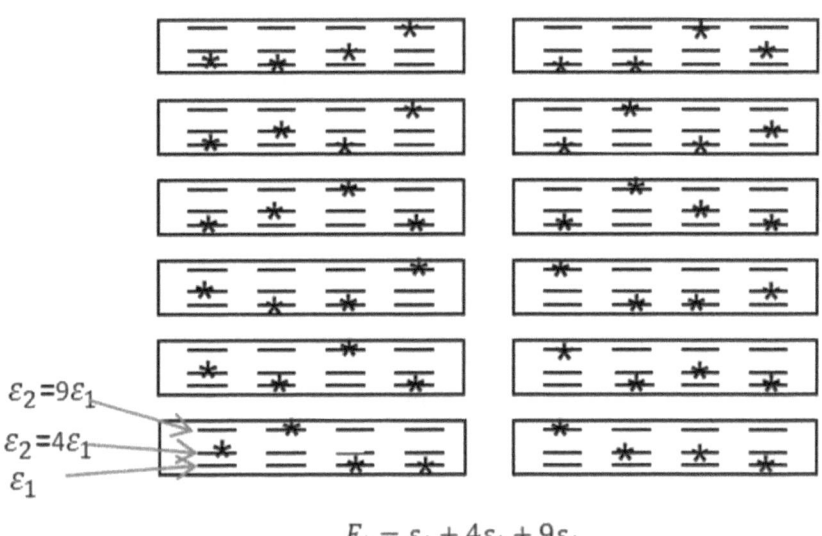

$$E_t = \varepsilon_1 + 4\varepsilon_1 + 9\varepsilon_1$$

Figure 2-1 Distribution of 4 particles with equal probability between three available energy levels.

The entropy of the system is related to the number of ways that the particles are distributed with equal probability among accessible energy levels at a given temperature and volume (in this one dimensional case "L").

The entropy could be easily calculated from the following formula originating from statistical thermodynamics:

$$S = k_B \ln W$$

Where k_B is the Boltzmann constant, a bridge between macroscopic and microscopic properties of matter. In SI units $k_B = 1.3806488 \times 10^{-23} J/K$.

One can calculate the number of ways that the particles could be distributed with equal probability among accessible energy states. Let's show the number of particles that have the energy ε_1 with a_1, the number of particles that have energy ε_2 by a_2 and so forth. In the example above two molecules have the energy of ε_1, one has the energy of ε_2 and one has the energy of ε_3.

Therefore we have:

$a_1 + a_2 + a_3 = 4$, in general

$\sum_i a_i = A$, where A is the total number of molecules.

In addition we can write;

$E_t = 2\varepsilon_1 + \varepsilon_2 + \varepsilon_3$, or in general;

$$E_t = \sum_i a_i \varepsilon_i$$

Now by using the following statistical formula we can calculate the number of ways that 4 particles are distributed among these accessible energy levels.

$$W = \frac{A!}{\prod_1 a_i!}, \text{ and for the above case,}$$

$$W = \frac{1 \times 2 \times 3 \times 4}{(1 \times 2) \times 1 \times 1} = 12$$

As the length of the potential well increases, the energy levels become closer to each other and the number of accessible energy levels, and therefore entropy of the system increases. Thus entropy is a function of volume (lengths or area for one or two dimensional cases). In addition the accessibility of states depends on the availability of energy, which depends on temperature.

Let's consider another case where the total energy of system $E_t = 4\varepsilon_1$, that is only energy for level 1 is available, then there exist only one distribution (case B).

For the case that total energy of the system $E_t = 30\varepsilon_1$, then one particle has the energy of ε_1, the second has the energy $4\varepsilon_1$, the third $9\varepsilon_1$ and the fourth $16\varepsilon_1$. There are 24 ways that these 4 particles could be distributed with equal probability among these four energy states (case C).

$$W = \frac{1 \times 2 \times 3 \times 4}{(1 \times 1 \times 1 \times 1)} = 24$$

Let examine the above three cases from another point of view. First, notice that certain conditions are imposed in all three cases (as any other system). In case B each particle is described by state function φ_1 and each one has the energy ε_1. Each particle has kinetic energy ½mv² and they move around within the boundaries imposed by potential energy, and they collide with each other. If the restriction of case B is removed, and restriction of case A is imposed, then although under this new condition the distribution of case B still are available but the particles will choose distribution of case A. If the condition of case C is imposed, the particles will choose the distribution of case C. In each case the system is in equilibrium state regardless of the fact that the particles collide with each other (elastic collision in this case) and kinetic energy is transferred from one particle to the other. Yet, in each case the system is in its equilibrium state, that is, if forcefully the system is brought to another state, once the imposing condition is removed, it will return to its equilibrium state.

Now we are in the position to review other interpretations of entropy, disorder being the most popular one. In case B there is only one way that particles could be distributed among energy levels under the imposed condition, therefore in this sense (one distribution), we have an ordered arrangement. In case A there are 12 ways, In case C there are 24 ways for distributing the particle among available energy levels, one can call this disorder and one call it justices. Since the energy of the universe is constant but its entropy increase, then on the disorder interpretations of entropy one can conclude that violence and disorder will govern the world and a rescuer is need to save the world or based on justice interpretations of entropy one comes to conclusion the universe goes to its final state of equilibrium.

Entropy sometimes is interpreted as information, the more information, less entropy. In case B, we know the energy of each particle is ε_1. In case A our information is limited we know that two of the particles have energy ε_1, one has energy ε_2 and the other one has the energy ε_3. Since our knowledge about the universe increase, then it is argued that knowledge and entropy co ntradict each other.

Also it has been argued that evolution contradicts the maximum entropy postulate. The question has been how the organization of living system could have raisin spontaneously? To resolve this apparent contradiction we explore the deep meaning of equation 3-9a. This equation means that the entropy of the universe increases spontaneously, but universe is a composite system of infinite numbers of subsystems. Living systems are part of this universe. So, the decrease in entropy is compensated by entropy increase in the other parts of the universe.

Surface thermodynamic

At the beginning of this chapter, we limited our discussion of thermodynamics to simple systems, which was described by three independent extensive parameters (U,V,n) in entropy representation or (S,V,n) in energy representation and ignored the surface effects. We also stated that the restriction of the theory to three extensive parameters does not limit the generality of the theory. Surface when is not negligible with respect to the volume, plays very important role in physical and chemical properties of the matter. The environment around surface molecules or atoms is different from those in the bulk, and for very fine material the surface effects are very significant. For example if a 1 mm solid cube is divided into 0.1 mm solid cubes, the surface increases 60 times.

To take the thermodynamic properties into account we add the extensive parameter "A" representing the surface of the system into the fundamental equations. Then, for one component system,

$$S = U(U,V,A,n)$$

Or in energy representation,

$$U = U(S, V, A, n)$$

In differential form,

$$dU = TdS - PdV + \gamma dA + \mu dn$$

Where,

$$\left(\frac{\partial U}{\partial A}\right)_{S,n,V} \equiv \gamma : \textit{The surface tension}$$

As pressure is defined as force to the area over which force is distributed; surface tension is defined as force per unit length of the surface that opposes the expansion of surface area.

The Gibbs free energy (equation 3-17) becomes,

$$dG = -SdT + VdP + \gamma dA + \mu dn$$

Chapter Three Supplementary

Intensive parameters

The Thermodynamic definition of temperature, pressure and chemical potential

The fundamental relation in the form of 3-5 or 3-6 if known contains all thermodynamic information about the system, but from thermodynamics one cannot deduce this fundamental relation. Thermodynamics gives information about the change from one equilibrium state to another. Therefore, the differential form of the fundamental relation plays the leading role in thermodynamics. If we write the fundamental relation for one component system in the form of energy,

$$U = U(S,V,n)$$

And compute the first derivative:

$$dU = \left(\frac{\partial U}{\partial S}\right)_{V,n} dS + \left(\frac{\partial U}{\partial V}\right)_{S,n} dV + \left(\frac{\partial U}{\partial n}\right)_{S,V} dn \qquad \text{S 3-1}$$

These partial derivatives are called intensive parameters and like extensive parameters they describe the state of system. The following notations are used for these partial derivatives.

$$\left(\frac{\partial U}{\partial S}\right)_{V,n} \equiv T : \text{The Temperature} \qquad \text{S 3-2}$$

$$\left(\frac{\partial U}{\partial V}\right)_{S,n} \equiv -P : \text{The Pressure} \qquad \text{S 3-3}$$

$$\left(\frac{\partial U}{\partial n}\right)_{S,V} \equiv \mu : \text{The electrochemical Potential (Chemical potential)} \qquad \text{S 3-4}$$

With these notations equation S 3-1 becomes,

$$dU = TdS - PdV + \mu dn \qquad \text{S 3-5}$$

or in terms of entropy,

$$dS = \frac{1}{T}dU + \frac{P}{T}dV - \frac{\mu}{T}dn \qquad \text{S-3-6}$$

For r component system,

$$dU = TdS - Pdv + \sum_{k=1}^{r} \mu_k dn_k \qquad \text{S 3-7}$$

Where, μ_k represents the chemical potential of the kth component.

Formal definition of temperature

Suppose that two subsystems are separated by an adiabatic and fixed wall which is impermeable to flow of matter. Furthermore assume that the temperatures of the two subsystems are almost, but not quite equal. Let's assume that,

$$T^{(1)} > T^{(2)}$$

The composite system is in equilibrium with respect to the imposed constraint. Now, if the adiabatic constraint is removed (that is the adiabatic wall is replaced by a diathermal wall), then the system no longer is in equilibrium, heat flows across the diathermal wall, and the entropy of composite system increases. Finally the system comes to a new equilibrium state determined by the condition that the final values of $T^{(1)}$ and $T^{(2)}$ are equal and entropy has its maximum value that is consistent with the remaining constraints. The energy of the composite system is constant, that is;

$$U^{(1)} + U^{(2)} = Constant \qquad \text{S 3-8}$$

But $U^{(1)}$ and $U^{(2)}$ will change because of heat flow between the two subsystems. Since

$$S = S^{(1)} + S^{(2)}$$

And

$$dS = dS^{(1)} + dS^{(2)}$$

Then, by using equation S 3-6 for constant volume and constant n, we have

$$dS = \frac{1}{T^{(1)}} dU^{(1)} + \frac{1}{T^{(2)}} dU^{(2)}$$

From the principal of energy conservation (equation S 3-8)

$$dU^{(2)} = -dU^{(1)}$$

Then,

$$dS = \left[\frac{1}{T^{(1)}} - \frac{1}{T^{(2)}}\right] dU^{(1)}$$

At the new equilibrium state entropy is maximum, that is $dS = 0$ and consequently

$$\frac{1}{T^{(1)}} = \frac{1}{T^{(2)}}$$

or

$$T^{(1)} = T^{(2)}$$

This means that if two systems with different temperatures are separated by a diathermal wall, heat will flow between two subsystems until their temperatures become equal. The entropy change between the initial equilibrium state and final equilibrium state is positive, thus

$$dS = \left[\frac{1}{T^{(1)}} - \frac{1}{T^{(2)}}\right] dU^{(1)} > 0$$

By the condition, $T^{(1)} > T^{(2)}$ now we have

$$dU^{(1)} < 0$$

This means that in this spontaneous process heat flow from high temperature subsystem (1) to low temperature subsystem (2).

The intuitive definition of temperature is based on the physiological sensation of hot and cold. We expect that temperature to be an intensive property, we also expect that heat to flow from high temperature region to low temperature region and at equilibrium the temperature be equal everywhere. These expectations are in agreement with the formal thermodynamic definition of temperature. The temperature is absolute Kelvin temperature designated by °K which is obtained by assigning the number 273.16 to a mixture of ice, water and water vapor in mutual equilibrium.

Mechanical equilibrium

Again consider a composite system made of two subsystems that are separated by an adiabatic and fixed wall which is impermeable to flow of matter. Now we replace the adiabatic and fixed wall separating the two subsystems with a diathermal and moveable wall. This means that energy and the volume of the two subsystems can change under the following conditions.

$$U^{(1)} + U^{(2)} = Constant$$

And

$$V^{(1)} + V^{(2)} = Constant$$

At the new equilibrium under the constraint of constant number of moles for each subsystem, the entropy of composite system is maximum, so $dS = 0$

From S 3-5 for constant number of moles we have,

$$dS = \frac{1}{T^{(1)}} dU^{(1)} + \frac{P^{(1)}}{T^{(1)}} dV^{(1)} + \frac{1}{T^{(2)}} dU^{(2)} + \frac{P^{(2)}}{T^{(2)}} dV^{(2)}$$

Since,

$$dU^{(2)} = -dU^{(1)}$$

And

$$dV^{(2)} = -dV^{(1)}$$

then,

$$dS = \left(\frac{1}{T^{(1)}} - \frac{1}{T^{(2)}}\right)dU^{(1)} + \left(\frac{P^{(1)}}{T^{(1)}} - \frac{P^{(2)}}{T^{(2)}}\right)dV^{(1)}$$

At the new equilibrium $dS = 0$, and,

$$T^{(1)} = T^{(2)}$$

And

$$P^{(1)} = P^{(2)}$$

The equality of pressure in the new equilibrium state is the result of movable wall.

Equilibrium with respect to matter flow

Consider the equilibrium state of a two simple systems separated by a rigid diathermal and impermeable wall. Now, if we replace the impermeable wall with a wall that is permeable to one component and impermeable to others, the system will reach to a new equilibrium state. The change of entropy is:

$$dS = \frac{1}{T^{(1)}} dU^{(1)} - \frac{\mu_1^{(1)}}{T^{(1)}} dn_1^{(1)} + \frac{1}{T^{(2)}} dU^{(2)} - \frac{\mu_1^{(1)}}{T^{(1)}} dn_1^{(1)}$$

Since

$$dU^{(2)} = -dU^{(1)}$$

And

$$dn_1^{(1)} = -dn_1^{(2)}$$

Then,

$$dS = \left(\frac{1}{T^{(1)}} - \frac{1}{T^{(2)}}\right) dU^{(1)} + \left(\frac{\mu_1^{(1)}}{T^{(1)}} - \frac{\mu_1^{(2)}}{T^{(1)}}\right) dn_1^{(1)}$$

At new equilibrium $dS = 0$, therefore

$$\mu_1^{(1)} = \mu_1^{(2)}$$

Just as temperature interpreted as a potential for heat flow, pressure as potential for volume change, chemical potential can viewed as potential for mater flow. The direction of mater flow can be analyzed by the same manner used for heat flow. If we assume that the above system was in thermal equilibrium, then

$$dS = \left(\frac{\mu_1^{(1)}}{T^{(1)}} - \frac{\mu_1^{(2)}}{T^{(1)}}\right) dn_1^{(1)}$$

If $\mu_1^{(1)} = \mu_1^{(2)}$, then $dn_1^{(1)}$ hould be negative since dS is positive and matter tends to flow from a high region of chemical potential to a lower region of chemical potential.

Reversible and irreversible process

A spontaneous process can be used as a source of energy for delivering work to an external system. Here we define an ideal external source of work in a way that the system can deliver work or receive work from this work source in a reversible manner. We also define an ideal external source of heat in a way that the system can deliver heat or absorb heat from this heat source in a reversible manner.

A reversible work source is defined as a system enclosed by an adiabatic impermeable wall and all the processes within the source are quasi static. The adiabatic wall insures that the entropy of this system is constant. The only flux of energy to or from a reversible work source is in the form of work.

A reversible heat source is defined as a system with a rigid impermeable wall and all the processes within the source are quasi static. The only flux of energy to or from a reversible heat source is in the form of heat. That is, $dU = đQ = TdS$

A very large reversible work or heat source (universe) is called a work reservoir or heat reservoir.

Maximum work processes

Consider the following composite system;

A subsystem in initial equilibrium state "A", that by some process attains the new equilibrium state, A reversible work source, and A reversible heat source as shown in diagram below:

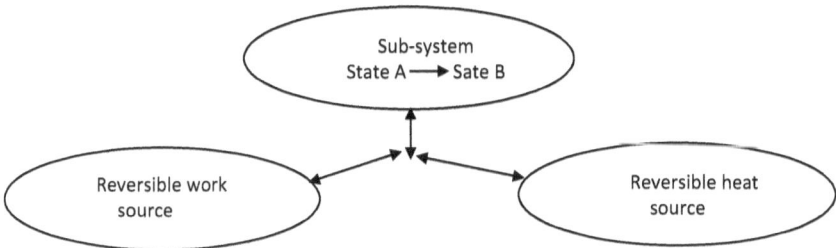

If the internal energy of the subsystem in state "B", U_B, is less than the internal energy U_A in state "A", then the extra energy is distributed between reversible work source and reversible heat source. We are interested to find the particular process that delivers the maximum amount of work to the reversible work source. The entropy of composite system in any real process increases, but the change of entropy in an ideal reversible process is zero, the Entropy change for the composite system and various potions of the composite system are given in the table 2-1.

Table 2-1 Comparison of entropy changes in reversible and irreversible process

System	Ideal reversible process	Irreversible process
Composite system	$\Delta S = \Delta S_{sys} + \Delta S_{rws} + \Delta S_{rhs} = 0$	$\Delta S = \Delta S_{sys} + \Delta S_{rws} + \Delta S_{rhs} > 0$
Sub-system	$\Delta S_{sys} = S_B - S_A$	$\Delta S_{sys} = S_B - S_A$
Reversible Work Source	$\Delta S_{rws} = 0$	$\Delta S_{rws} = 0$
Reversible Heat Source	$\Delta S_{rhs} = \Delta S - \Delta S_{sys} = \Delta S_{sys}$	$\Delta S_{rhs} = \Delta S - \Delta S_{sys}$

The entropy change for Reversible work source in ideal reversible process and any real process because of the surrounding adiabatic wall is zero. Therefore, entropy increase for reversible work source in ideal reversible process will be equal to the decrease of entropy of the subsystem, but in any real process because the entropy change in the composite system is a positive quantity, the entropy increase in reversible work source will be more. As the result less heat will be delivered to the reversible heat source in any real process than ideal reversible process and more work energy will be available. For a reverse process the amount of work from reversible work source will less for an ideal reversible process than any real process.

Thermodynamic efficiency is defined by;

$$\varepsilon = \frac{đW}{đQ} = \frac{đW'}{đQ'} \qquad \text{S 3-9}$$

Where, $đW'$ stands for work done by the system and $đQ'$ stands for the heat released by the system. Since in an ideal reversible process less heat is released by the system and more work is done by the system, then the efficiency for a reversible process will be maximum.

Alternative representation of thermodynamic fundamental relation

Legendre transformation

In both energy and entropy representation of fundamental relations, extensive parameters plays the role of independent parameters, and intensive parameters are introduced as the first derivative. However, it is much easier to measure and control the intensive parameters. It is possible by a mathematical transformation, known as Legendre Transformation, to transform the fundamental relation in such a way that temperature replaces entropy and pressure replaces volume.

For relation,

$$f = f(x, y)$$

We want to replace x with the first derivative of f with respect to x, That is we want to replace x with the slope of the curve at point (x, y). The slope is defined by:

$$\frac{Y\ b}{}, \text{ where } \rho \text{ is the slope and } b \text{ the y intercept is the Legendre transform of } Y$$

Solving for b, gives;

$$b = Y - \rho X$$

Replacing entropy with temperature leads to definition of Helmholtz free energy. Volume by pressure leads definition of enthalpy and replacing both entropy and pressure by temperature and pressure respectively leads to definition of Gibbs free energy.

Maxwell relations

At constant n four relations exist between the second derivatives of extensive parameters for one component systems.

$$\left(\frac{\partial T}{\partial V}\right)_S = -\left(\frac{\partial P}{\partial S}\right)_V \qquad \text{S 3-10}$$

$$\left(\frac{\partial V}{\partial S}\right)_P = \left(\frac{\partial T}{\partial P}\right)_S \qquad \text{S 3-11}$$

$$\left(\frac{\partial S}{\partial V}\right)_T = \left(\frac{\partial P}{\partial T}\right)_V \qquad \text{S 3-12}$$

$$\left(\frac{\partial S}{\partial P}\right)_T = -\left(\frac{\partial V}{\partial T}\right)_P \qquad \text{S 3-13}$$

Recall that change in the thermodynamic parameters with the exception of work and heat are independent of path, for example if one wants to calculate the change of entropy with respect to volume at constant temperature, then according to the relation S 3-12 the change will be equal to change of pressure with respect to temperature at constant volume which can be shown is equal to measurable quantities $\frac{\alpha}{\beta}$.

Chapter Four

Reactions

In order for a reaction to happen between species certain conditions should exist. First the species should be ready for the reaction and they should meet each other properly. The first part of this condition for a chemical reaction is dictated by the potential energy wells of involved species and in the case of humans, by their behavior wells. However, in the case of interaction between people the time dependency of behavioral well should be taken into consideration.

As an example, for the reaction between $H_2(g)$ and $Br_2(g)$, the latter has a deep potential well and the former has a broader and shallow potential well. The vibrational and rotational energy levels in the former are further apart than the latter. So it is quite possible that the initiation step in the above reaction to be,

$Br_2 + M \rightarrow 2Br + M$

M is the body that molecule upon collision with, acquires enough energy to overcome the potential energy and dissociate. Generally, absorption of heat from the walls as well as collision with other molecules is the source of energy. The species Br called a radical, is unstable species and seeks a partner. If it meets another Br radical, then the following reaction takes place:

$Br + Br \rightarrow Br_2$

In order for Br radical to interact with $H-H$ and produce HBr, Br should approach $H-H$ in a path that requires the least energy. In this case it would be along the line connecting two hydrogen atom. For a rather simple reaction $C + AB \rightarrow AB + C$ where the goal of C is to steel A from B the most effective approach will be along the line connecting A to B, but opposite to B. For a larger molecule the best course of approach will be from less hindered path. In addition the Br radical in our simple example or in general C should have enough energy to perturb the potential well of H_2, (AB). As Br gets closer to reacting H, the widths of potential well for H_2 increase, the rotational and vibrational energy levels of H_2 get closer to each other and the influence of Br on reacting H increases and the influence of the two hydrogen atoms on each other decreases. At some point the potential energy surface for $Br \mathrel{\text{L}} H \mathrel{\text{L}} H$ is minimum in all direction and maximum in one direction. Such a unique state is called transition state, and because the potential energy surface at this point is similar to a saddle, the transition state sometimes is called saddle point, and the curve that connects all points of minimum energy is called reaction path or reaction coordinates.

The $Br \mathrel{\text{L}} H \mathrel{\text{L}} H$ species, called activated complex and in the general form is shown by $[ABC]^{\ddagger}$ is like a normal molecule with the exception that one of its vibrational frequency is imaginary implying a negative force constant which means that in one direction in nuclear configuration space the energy is maximum, while in all other (orthogonal) directions the energy is minimum. Simply it implies that the activated complex either moves forward to produce HBr and H radical or backward to H_2 and Br. The ratio of forward process to backward process depends on the depth and widths of HBr and H_2. The active hydrogen radical can approach $Br - Br$ and via the activated complex HL BrL Br produce HBr and Br or react with another H radical to produce H_2.

In classical description of the above reaction the approaching molecule should have enough kinetic energy called activation energy, to overcome the potential energy holding the two hydrogen atom together.

The above description of interaction between two species resembles the interaction between people, how a relationship is formed and how is broken in terms of their behavioral well. In the above example the species approach each other in gas phase like people passing each other in street, where the activation energy for interaction usually is high. However, if there was a match maker then an alternative route for interacting would be available. In chemical reaction a catalyst plays the role of match maker and provides an alternative route with lower activation energy for the reaction.

The absolute rate theory of Eyring[6] and advanced computational in easy to use software such as Gaussian has enabled the chemist and biochemists to study the potential energy surfaces between equilibrium state of the reactant, transition state and equilibrium state of the reactant. In quantum chemistry calculations, the reaction path could be found first by finding the optimized structure of the reactants, product and transition state, and then by IRC command in the Gaussian software. IRC stands for Intrinsic Reaction Coordinate in mass weighted Cartesian coordinates.

---

[6] The Activated Complex in Chemical Reactions. H. Eyring, The Journal of Chemical Physics 3 (2), 107-115, 1935. For historical review see Development of Transition state theory. K. J. Laidler and M.C. King, Journal of Physical Chemistry **87** (15), 2657-2664 and Quantum and Semi classical Theory of Chemical Reaction rate, W.H. Miller, Faraday Discussions, 110, 1-21.

Rate of Reaction

Rate of reaction, that is the number of *HBr* molecules produced in unite time, for the general reaction $A + B = C + D$ is given by;

$$Rate = k[A]^{\alpha}[B]^{\beta}[C]^{\gamma}[D]^{\delta}$$

Where *k* is the rate constant, a temperature dependent constant, [ ] represent the number of molecule (moles) per unit volume or simply concentration, α, β, γ and δ are the order of reaction for species A, B, C and D respectively which are determined by experimental techniques.

For our example reaction in the early stages when the concentration *HBr* is negligible the experimental rate of reaction is given by

$$Rate = k[H_2][Br_2]^{1/2}$$

That is the reaction is first order with respect to H_2, 1/2 order with respect to Br_2 and zero order with respect to *HBr*. As the concentration of *HBr* increases the expression becomes complex

$$Rate = \frac{k[H_2][Br_2]^{1/2}}{1 + k'[HBr]/[Br_2]}$$

Both of these reaction rates shows that the $H_2 + Br_2$ reaction, is not a single step reaction but involves a series of reactions called elementary reactions as discussed above.

Thermodynamics of Reaction

Thermodynamics deals with the initial and final equilibrium states, that is, with the initial state of $H_2 + Br_2$ and final state of $2HBr$. Since the internal energy, entropy and other thermodynamic potentials are state functions, then the change for entropy and thermodynamic potentials are independent of the above elementary reactions. Entropy according to the third law of thermodynamic (postulate III, chapter three) is zero at absolute zero and at any temperature and pressure it can be calculated with reference to this point.

Again it should be pointed out that change in thermodynamic parameters are independent of path, that is a chemical reactions could start at a given temperature and pressure (or temperature and volume) and end up at another temperature and pressure (or temperature and volume), but for calculation purposes we can use stepwise calculations.

In classical thermodynamics a reference point is defined for enthalpy instead of energy. Since enthalpy and energy are related, then the energy can be calculated with respect to this reference point. It should be emphasized that the change in thermodynamic potentials and entropy will be independent of the choice of the reference point provided that the same standard is used for the species involved in that process.

Enthalpy of an element in its most stable form at 25 °C and standard pressure (1 bar) is chosen to be 0, thus enthalpy of H_2 (gas) and Br_2 (liquid) will be zero at this standard reference point. The enthalpy of formation for any molecule is defined as the heat of a reaction on which the molecule in its most stable form at 1 bar and 298.15 K is formed from its constituent elements at their most stable forms at 1 bar and 298.15 K. The heat of reaction is measured by an instrument called Calorimeter. However, for most reaction such as the above reaction it is not possible to perform the reaction in a Calorimeter in a manner that only the product of interest is produced. In these cases the heat of reaction is obtained indirectly by employing the fact that the change in thermodynamic properties is independent of the path. Here we will not get involved with the method that heat of reaction for the above reaction was measured. The heat of formation for any molecule can be found in standard table for thermodynamic values.

The heat of formation for $HBr(g)$ at standard pressure and temperature, calculated by indirect method is -36.40 kJmol^{-1}. Since energy and enthalpy are related by $H = E + PV$; then $\Delta H = \Delta E + \Delta PV$

If we assume that the gases involved in this reaction have ideal behavior, then:

$\Delta PV = RT \Delta n$

Where Δn refers to the net change of moles of gases involved in the reaction; in this case

$\Delta n = 2\,mole\ HBr(g)\ produced - 1\,mole\ H_2(g)\ consumed = 1$

Therefore, $\Delta E = \Delta H - RT$, or

$\Delta E^0_{298.15} = -36.40 kJmol^{-1} - 8.314 JK^{-1}mol^{-1} \times$

$298.15 K \times 1kJ\!\!\diagup\!\!\!_{1000 J} = -38.88 kJmol^{-1}$

According to the third law of thermodynamics the entropy of any molecule at zero Kelvin in a perfect crystal is zero. Then, entropy of any molecule at any temperature can be calculated by employing the proper thermodynamic formulas.

Entropy of $H_2(g)$; $Br_2(l)$ and $HBr(g)$ at standard pressure of 1 bar and temperature of $298.15 K$ are $130.684, 152.231$ and $198.695\ JK^{-1}mol^{-1}$ respectively. Therefore:

$$\Delta S^0_{298.15} = 2mol \times 198.695 JK^{-1} - (1mol \times 130.68 JK^{-1}mol^{-1} + 1mol$$

$$\times 152.23 JK^{-1}mol^{-1}) = 114.475 JK^{-1}$$

Or change in entropy per mole at 298.15 K and standard pressure is $57.238 JK^{-1}mol^{-1}$.

Since $G = H - TS$, therefore at constant pressure and temperature.

$$\Delta G^0_{298.15} = -36.40 kJmol^{-1} - 298.5 K \times 57.238 JK^{-1}mol^{-1} \times$$

$$\frac{1 kJ}{1000 J} = -53.47 kJmol^{-1}$$

The negative value of ΔG indicates that this reaction at standard pressure and at 298.15 K is allowed by thermodynamic laws. Again it should be emphasized that thermodynamics is only concerned about the change.

At the molecular level the reference point for the energy of molecule is defined by the separate atoms, and the energy of an atom is defined with respect to separated electrons and nucleus. The energy for any atom or molecule can be calculated by solving equation

$$\hat{H}\psi_q = E\psi_q$$

For Hydrogen atom this equation can be solved easily, but for heavier atoms and molecules this equation is solved by approximation methods. The theoretical results at two level of calculations, B3lyp/6-311G* and MP2/6-311G** using Gaussian Package are compared with thermodynamic results in the table 4-1.

Table 4-1 Theoretical and experimental results for $H_{2(g)} + Br_{2(g)} = 2HBr_{(gas)}$ reaction.

	Theoretical		Thermodynamics
	B3LYP/6-31G*	MP2/6-311G**	
$\Delta E^0_{298.15} kJmol^{-1}$	-105.489	-120.201	-111.678
$\Delta H^0_{298.15} kJmol^{-1}$	-103.011	-117.722	-109.200
$\Delta S^0_{298.15} Jmol^{-1}K^{-1}$	21.326	21.213	21.243
$\Delta G^0_{298.15} kJmol^{-1}$	-124.08	-109.336	-115.534

Since in the above theoretical calculations, molecules are in their gas phase state for comparison we calculate these parameters for the reaction

$H_2(g) + Br_2(g) = 2HBr(g)$ at 1 bar pressure and 298.15 K, the change in thermodynamic parameters

$$\Delta S^0_{298.15} = 2mol \times 198.695 JK^{-1}mol^{-1} - 1mol(130.68 + 245.463)JK^{-1}mol^{-1}$$

$$= 21.243 JK^{-1}$$

$$\Delta H^0_{298.15} = 2mol \times (-30.907)kJmol^{-1} - 1mol \times (0 + 36.400)kJmol^{-1}$$

$$= -109.200 kJ$$

$$\Delta E^0_{298.15} = -109.200 kJmol^{-1} - 8.314 JK^{-1}mol^{-1} \times \frac{1kJ}{1000J}$$

$$= 111.678 kJmol^{-1}$$

$$\Delta G^0_{298.15} = \Delta H^0_{298.15} - T\Delta S^0_{298.15}$$

$$= -109.200 KJ - 298.15k \times 21.243 JK^{-1} \times \frac{1kJ}{1000J}$$

$$= 115.534 kJmol^{-1}$$

Again it should be emphasized that constant temperature and pressure does not mean that these two parameters are necessarily constant during the reaction course, it simply means that the temperature and pressure of the products at final state will be the same as the initial stage. Negative values for $\Delta G^0_{298.15}$ indicates the above reaction will take place at 1 bar pressure and room pressure, however, this value does not provide any information about the rate of reaction.

The change in Gibbs free energy is related to equilibrium constant for any reaction by

$$\Delta G^0 = -RT\ln K_{eq}$$

Where, K_{eq} the equilibrium constant is related to the activity of the each species involved in the chemical reaction:

$$K_{eq} = \prod_i a_i^{v_i}$$

Since ΔG^0 is independent of pressure then equilibrium constant depends only on temperature.

In most cases to simplify the calculation one replaces the activity with partial pressure, then

$$\Delta G^0 = -RT\ln K_P$$

Where,

$$K_P = \prod_i P_i^{v_i}$$

For our example reaction,

$$K_P = \frac{P_{HBr}^2}{P_{H_2}}$$

Where the reference point for pressure is 1 bar and partial pressure of liquid Br_2 (or any liquid or solid) is taken to be 1.

Reaction at Cold interstellar space

In classical treatment of chemical reactions, the rate of reaction depends on activation energy, a temperature dependent parameter. In cold space there is not enough energy to overcome the activation energy, so the chemical reactions should slow down as the temperature decreases, and even ceases at very low temperature in space. Yet, molecules such as hydrogen and methanol are abundant in dense, cold, interstellar clouds. Methoxy radical (CH_3), methylidyne radical (CH), polymeric composition based on the formaldehyde molecule

(H_2CO), polycyclic aromatic hydrocarbons (PAHs) and more are detected.

Formation of hydrogen molecule in abundance in gas phase cannot be explained by H^- route because of very low densities of free electrons in dense cloud.

$H + e \rightarrow H^- + h\nu$

$H + H^- \rightarrow H_2 + e$

Formation of H_2 is explained by interstellar dust grain mechanism. The hydrogen atoms are adsorbed onto the surfaces of grains and concentrated enough for two atoms to react. The reaction $H\left(over\ grain\right) + H\left(over\ grain\right) \rightarrow H_2\left(over\ grain\right)$ is very exothermic and liberates enough energy to liberate H_2 from the surface of grains. This mechanism has been verified experimentally[7]

The methoxy molecule $\left(CH_3O^-\right)$ is found in Perseus Molecular Cloud. The existence of this radical is attributed to the reaction between methanol (CH_3OH) and hydroxyl radical $\left(OH\right)$. Reproduction of this molecule by allowing the reactants condense on dust grain has not been successful.

A team of scientists at the University of Leeds, UK, recreated the cold environment of space in the laboratory and studied the reaction

---

[7] Analysis of Molecular Hydrogen formation on Low Temperature Surface in Temperature Programmed Desorption Experiments. G. Vidali *et al*. J. Phys. Chem. A, 2007, 111(49), 12611.

of methanol with hydroxyl radical at minus 210 degrees Celsius[8]. They found that not only do these gases react to create methoxy radicals at such an incredibly cold temperature, but that the rate of reaction is 50 times faster than at room temperature and this faster reaction can only occur in gas phase in space.

The interaction of OH radical with CH_3OH at so low temperature is explainable by quantum tunneling through the barrier. But why the reaction is so much faster than room temperature needs more justification. According to professor Heard, head of research team at Leads, the cold slows down the molecules, which allows the molecules to stay close to each other instead of quickly bouncing, and increase the opportunity for tunneling.

---

[8] Accelerated chemistry in the reaction between the hydroxyl radical and methanol at interstellar temperatures facilitated by tunnelling. R.J. Shannon, M.A. Blitz, A. Goddard and D.E. Heard. Nature Chemistry **5**, 745-749, 2013. Low Temperature Kinetics of CH_3OH+OH reaction. C. Gómez Martín, R. L. Caravan, M. A. Blitz, D. E. Heard, and J. M. C. Plane. J. Phys. Chem. A,**2014**, 118(15), pp 2693-2701

Chapter Five

Applications

In this chapter we will apply the theories discussed in previous chapters to a simple molecule, water, and its components hydrogen and oxygen.

Water is a simple, but a mysterious molecule. It is the only substance that exists abundantly in liquid, solid and gas physical states. It is found everywhere from interstellar dust clouds to orange-red fields of Mars mostly in the form of ice. Water at very high temperature and pressure exists in plasma state (charged nucleolus and free electrons). Water in the plasma state has been detected in the interior of planet Glies 1214 b.

Scientists are looking for clues to find out how water was collected on Earth billions years ago. Water vapor formation on young hot star such as Orion-BN-K is of great interest to scientists. Hot-young stars through off powerful winds and jet streams which upon collision with materials such as oxygen, generates shock waves that forces oxygen to react with hydrogen, which is abundant in interstellar cloud[9].

---

[9] Under the action of shockwave the velocity of reactants can reach 1000m/s, the same order as the vibrational velocity of atoms in molecules. This induces chemical reaction directly.

The absorption coefficient of water in the visible region of the spectrum is one million times lower than the rest of the spectrum, thus the water vapor in the atmosphere allows the visible light from the sun and reflected light from earth to pass. Absorption of infrared light by water molecules in the atmosphere keeps the earth warm.

Because of three dimensional hydrogen bonding, water has high melting and boiling points. It has a high heat of fusion, a high heat of sublimation, a high heat of vaporization, and a high heat capacity and a high surface tension; all of these properties are related to the extra energy needed to break the hydrogen bond. These properties make water a unique substance that plays a vital role in life.

A- Historical Aspects

In 18th century it was discovered that water is made from oxygen atom and hydrogen atoms, and it was well established, that each molecule of water is made of two hydrogen atoms and one oxygen atom. In order to explain the properties of water, we should understand the properties of hydrogen and oxygen.

Hydrogen gas was first produced in the early 16th century by adding acids to metals. In 1766–81, Henry Cavendish was the first to recognize that hydrogen gas was a discrete substance, and that it produces water when burned. Thus, later this substance was called hydrogen, in Greek, hydrogen means "water-former".

Most of Earth's hydrogen exists as diatomic gas H_2. Hydrogen atoms rarely exist in Earth's atmosphere, about 1ppm by volume.

Hydrogen is the most abundant element in the universe, and mostly exists in atomic and plasma states. As plasma, hydrogen's electron and proton are not bounded together, as the result hydrogen in plasma state has a very high electrical conductivity and high emissivity (the source of light from the sun and other stars). The charged particles are highly influenced by magnetic and electric fields. For example, in the solar wind they interact with magnetosphere giving rise to aurora (a natural light in the sky more common to higher latitudes). Hydrogen in neutral atomic state is found in the interstellar medium.

A molecular form of hydrogen called protonated molecular hydrogen H_3^+ is found in the interstellar medium where is generated by ionization of molecular hydrogen by cosmic rays. This charged molecule has also been observed in the upper atmosphere of the planet Jupiter. The ion is relatively stable in the environment of outer space due to the low temperature and density of the said environment.

The infrared interstellar bands are attributed to $H-H^+$ and $H-H$ collisions. Ab initio calculations of hydrogen molecule at excited electronic states has the potential of providing theoretical insight to the origin of these bands[10]

B- Theoretical Aspects- Structure and Electronic Configuration

Hydrogen atom

Hydrogen atom consists of one proton and one electron, and the interaction between them which depend on the position of electron with respect to proton. It should be noted, however, that proton is 1831 times heavier than electron. So by choosing the proton as the origin of the coordinate system the problem reduces to one. Yet

---

[10] See for example: Ab Initio calculation of molecular hydrogen electronic states, Properties, Transition matrix elements among triple electronic states . A. Spielfiedle, P. Palmieri and Mitruhenkov; Molecular physics, V 102, Pages 2249-2258

in Cartesian coordinates the potential energy between proton and electron $V_{(x,y,z)}$ depends on 3 coordinates. Therefore the separation technique cannot be used as in the case as of two dimensional particle in this coordinate system. However, in spherical coordinate system defined by (r,θ,φ), potential depends only on r. Now the separation of variables techniques can be used to reduce the three dimensional problem to three one dimensional problem. The process, however, is beyond the scope of this book, so we will discuss the results in a qualitative manner.

$$\Psi_{(r,\theta,\varphi)} = R_{(r)} \grave{\mathrm{E}}_{(\theta)} \phi_{(\varphi)}$$

Solution to $R_{(r)}$ give rise to a radial function which depends on an integer represented by n called the principal quantum number, ranging from one to infinity. Energy depends on this quantum number. Solution to polar part $\Theta_{(\theta)}$ leads to a polar function, which depends on quantum number l called orbital quantum number. Values of l are restricted to $1, 2, 3 \ldots n-1$. The solution to azimuthal part $\phi_{(\varphi)}$ depends on quantum number m. Values of m are restricted to $m_l = -l, -l+1, \ldots +l$.

Spin function and spin quantum numbers

Spin quantum numbers for electron (and thus for all elementary particles) was proposed to explain the experimental observations. Spin somehow is associated with spin angular momentum (no counterpart in classical mechanic). Spin function was an essential part of relativistic theory derived by Paul Dirac in 1928.

For electrons there are only two spin functions, which in matrix description are denoted by $|\alpha\rangle$ and $|\beta\rangle$ with the respective

Eigen-values of ½ h and −½h. In natural units h is eliminated and spin quantum numbers for electrons simply are +1/2 and -1/2.

We have ignored the mathematical presentation of the functions describing the electron in hydrogen atom at different energy levels because they are not suitable for presentation for this journey. However, they are the basis for understanding the behavior of electrons in atoms with more than one electron. So from now on we will use the following symbols for these functions.

All functions corresponding to $l = 0$ are designated by letter s. For the function corresponding to $n = 1$, the designated symbol is shown by $1s$, for $n = 2$ by $2s$ and so forth.

All functions corresponding to $l = 1$, are designated by letter p. Considering the radial part of the equation that is the functions corresponding to n, they are shown by $2p$, $3p$ and so forth. Notice that there is no $1p$ because for $n = 1$, $l = 0$.

All functions corresponding to $l = 2$ are shown by $3d$, $4d$ and so forth

All functions corresponding to $l = 3$ are shown by $4f$, $5f$ and so forth.

Recall that in the case of particle in two dimensional space with $L_x = L_y$, there were two distinct functions with the same energy and they were called degenerate states. Degeneracy will be removed if a potential is applied in one direction. In the case of hydrogen atom energy depends only on quantum number n and not on l, m, and spin functions. So for the case of $n = 2$ and $l = 0$ ($2s$) and $n = 2, l = 1, m_l = -1, 0, 1$ ($2p$) altogether there are 4 functions with the same energy, so the degeneracy is 4. We have ignored spin part of the equation. For the case of $n = 3$ degeneracy is 9. If another electron is added to the system, for example in helium atom with two electrons due to interaction between the two electrons the energy for the $\Psi_{(2,0,0)}$ function will be different from the $\Psi_{(2,1,m)}$ functions and the degeneracy reduces to 3. The energy for the $\Psi_{(3,1,0)}$ functions will be different for the $\Psi_{(3,2,m)}$ functions and degeneracy reduces to 3 and 5. To simplify the representations it is customary to call $\Psi_{(n,0,0)}$ functions ns orbitals, $\Psi_{(n,1,m)}, n>1$, np orbitals, and $\Psi_{(n,2,m)}, n>2, nd$ orbitals. I am reluctant to use the orbital terminology because they refer to classical concept rather than quantum concept. Nonetheless, the terminology is widely used by the chemists.

Oxygen atom

Besides hydrogen atom or precisely one electron systems such as H_2^+, multi electron quantum mechanics equations cannot be solved except by approximation methods.

In the simplest case, one ignores all interactions between electrons, so the problem reduces to n one electron systems. Thus one uses the hydrogen atom state functions called orbitals to describe the electronic state of multi electron atoms. With this approximation the error on binding energy for helium atom is 37.7% which reduces to 1.9% for the most rigorous calculations.

In the simplest case the ground electronic state of helium atom is represented by $1s(\alpha), 1s(\beta)$, where spin state (α) and (β), are used in this representation to show that according to Pauli exclusion principal for fermions (particles with half integer spin) two electrons with the same quantum state (orbitals) have different spins.

For the case of oxygen with eight electrons in the simplest case, the ground electronic state is represented by $1s(\alpha)1s(\beta)2s(\alpha)2s(\beta)2p_z(\alpha)2p_z(\beta)2p_x(\alpha)2p_y(\alpha)$ where in addition to Pauli Exclusion Principle; the Hund rule is applied here. The Hund rule is an observation rule which states the greatest total spin state usually makes the atom more stable. It is much simpler to show the electronic structure schematically. For oxygen atom:

$2p_z(\alpha)2p_z(\beta)$ ⥮ $2p_x(\alpha)$ ↑ $2p_y(\alpha)$ ↑

$2s(\alpha)2s(\beta)$ ⥮

$1s(\alpha)1s(\beta)$ ⥮

(Energy)

Where the energy scale is arbitrary and arrows up and down arbitrary represent spin state α and β. Also notice the degeneracy of $2s$ and $2p$ orbitals are removed because interaction between electrons qualitatively is taken into consideration.

It should be emphasized that above representation is only for an oxygen atom and two unpaired $2p$ electrons represents the paramagnetic properties of oxygen atom. When oxygen atom is bonded to another atom, s and p orbitals are not sufficient to describe the state of molecule, even qualitatively.

Hydrogen molecule

In the qualitative description of hydrogen molecule, the interaction of the two electrons is being ignored, and the ground electronic state is described by linear combination of two $1s$ orbitals:

$$\psi_+ = \frac{1}{\sqrt{2}}(1s_A + 1s_B)$$

and

$$\psi_- = \frac{1}{\sqrt{2}}(1s_A - 1s_B)$$

Where $1s_A$ and $1s_B$ refers to nucleus A and B. This means that the value of the functions at each point in space should be added "ψ_+" or subtracted "ψ_-". Then the first one will show a bonding state and the second one a non bonding state. Since molecular orbitals are obtained from linear combination of atomic orbital, the method is called MO-LCAO. $\frac{1}{\sqrt{2}}$ is called normalization constant to satisfy the requirement of a "well behaved" function.

For bonding state the spin function is:

$$\frac{1}{\sqrt{2}}[\alpha(1)\beta(2) - \alpha(2)\beta(1)]$$

And for nonbonding state:

$$\alpha(1)\alpha(2)$$

$$\beta(1)\beta(2)$$

$$\frac{1}{\sqrt{2}}[\alpha(1)\beta(2) + \alpha(2)\beta(1)]$$

Therefore the bonding state of hydrogen molecule is singlet and the non bonding state is triplet.

We should point out that in above description, the interaction between two electrons is not taken into consideration, and quantitative energy calculation is off by about 100%. On the best approximation using self consistent methods, the calculated binding energy for H_2 is an error by about 2%.

Alán Aspuru-Guzik, and his group at Harvard University have built a quantum computer for precise calculations."

It should be mentioned that in above treatments of molecules, the atoms preserve their identity. In Hardonic mechanics a strong attractive force between two electrons is introduced for short range of overlapping. Once the electrons are bonded they become Boson (with charge zero, spin 0 and magnetic moment 0 and radius of 6.843329×10^{-11} cm).

Para-Hydrogen and Ortho-Hydrogen

In hydrogen molecule the spin of the two protons couple to form a triplet state called ortho-hydrogen and a singlet state called para-hydrogen. The ratio of para to ortho depends on temperature.

Oxygen molecule

Here like hydrogen molecule, the atomic orbitals of oxygen atom at different levels of approximation are taken as the basis set for the calculation of molecular orbitals of oxygen molecule. In the most simplified calculation the atomic orbitals for hydrogen atom are chosen to represent the atomic orbitals of oxygen atom and consequently they are used for construction of molecular orbitals, as shown below.

$2p_z, 2p_x, 2p_y$

$\sigma_{2p}^* = \frac{1}{\sqrt{2}}(2p_z(A) - 2p_z(B))$

$\pi_{2p}^* = \frac{1}{\sqrt{2}}(2p_x(A) - 2p_x(B))$, $\frac{1}{\sqrt{2}}(2p_y(A) - 2p_y(B))$

$\pi_{2p} = \frac{1}{\sqrt{2}}(2p_x(A) + 2p_x(B))$, $\frac{1}{\sqrt{2}}(2p_y(A) + 2p_y(B))$

$2p_x, 2p_y, 2p_z$

$\sigma_{2p} = \frac{1}{\sqrt{2}}(2p_z(A) + 2p_z(B))$

$\sigma_{2s}^* = \frac{1}{\sqrt{2}}(2S(A) - 2S(B))$

$2s$ $2s$

$\sigma_{2s} = \frac{1}{\sqrt{2}}(2S(A) + 2S(B))$

Oxygen atom (A)   Oxygen molecule   Oxygen atom (B)

In the above qualitative representation of energy levels of oxygen molecule, the energy separation is arbitrary. The energy of σ_{2p}^* level is chosen to be higher than π_{2p}^* levels to predict magnetic properties of oxygen molecule.

Rotation-Vibration Energy Levels of Hydrogen Molecule and Oxygen Molecule.

Rotation-Rigid Rotor Approximation

In this approximation it is assumed that the distance between two atoms in diatomic molecule does not change. In a more rigorous treatment, the relative motion of the two atoms (vibration) should be taken into account.

$$\hat{H}_{u}\psi_{rot} = E_{rot}\psi_{rot}$$

In the case of diatomic molecule, there is two degree of freedom and the equation could be solved by separation of variable techniques. Solution is similar to the solution of polar part of hydrogen atom. Energy of rigid rotator is given by:

$$E_J = BJ(J+1)$$

Where $B = \dfrac{h^2}{2I}$, $I = \mu r^2$, $\mu(reduced\ mass) = \dfrac{m_1 m_2}{m_1 + m_2}$ and

$J_{(rotaional\ quantom\ number)} = 0, 1, 2, \ldots$

Selection rules only permits transition between consecutive energy levels and requires that the molecule to have a permanent dipole moment.

The energy difference between two successive energy levels is:

$$\Delta E = 2B(J+1)$$

Since the energy difference between rotational levels falls into microwave region (1-10 cm^{-1}) rotational spectroscopy is called microwave spectroscopy.

In rigid rotor approximation the energy difference between successive energy level is equal to $2B$, but the distance between two atoms due to vibration is not fixed, as the result the energy levels get closer as J increases.

Due to dipole moment requirement hydrogen molecule has no pure rotational spectra or as is called is not rotationally active.

Vibration- harmonic oscillator approximation

Near the bottom of potential well for diatomic molecules, the potential energy curve matches the potential curve for harmonic oscillator, thus to a first approximation vibrational energy levels is given by

$ü_v = \left(+ \frac{1}{2} \right) v$, in Joules units. In more common unit "wave number $\bar{v}$", $E_v = \left(n + \frac{1}{2} \right) \bar{v}$. Where, $\bar{v} = \frac{1}{2\pi}\sqrt{\frac{K}{\mu}}$, K, the force constant and μ is the reduce mass. In this approximation the difference between successive energy levels is equal to v in joule units and to $\bar{v}$ in cm^{-1}. when the anharmonicity is taken into consideration the energy levels get closer as n, vibration quantum number increases.

The transition between successive energy levels is allowed and, ΔE falls into infrared rejoin of spectrum. In order for a molecule to absorb a photon, its dipole moment should change with vibration; therefore, hydrogen molecule is not active in infrared. However, other types of spectroscopy such as Raman and Coherent Anti-Stock

Raman Spectroscopy (CARS) can provide information about the rotational and vibrational level of homo-nuclear diatomic molecules.

It is instructive to note that, homo-nuclear molecules such as N_2 and O_2 the major constituent of earth atmosphere cannot absorb the infrared radiation from the sun or emitted by earth. Whereas molecules such as CO_2, H_2O, NH_3 and CH_4 have electric dipole moment and absorb infrared radiation. Although these gases are present in small concentration, but they are responsible for keeping a delicate balance between sun energy and the energy emitted by earth. A small increase in amount of these gases can increase the temperature of earth.

Water molecule

The structure of water is being extensively re-investigated in universities and top research institutes around the globe. The first explanation of anomalous properties of water was qualitatively explained by W.K. Röntgen on 1892. According to Röntgen[11], liquid water consists of two types of molecules with different structure.

In 1933, after investigating X-ray diffraction data for water, Bernal and Fowler[12] proposed a theoretical model for water, where the oxygen was placed in the center of a tetrahedron and two hydrogen occupied the two vertices of tetrahedron and the two very negative pairs electrons

---

[11] See the review paper on "Centennial of Röntgen paper" on the structure of water, Journal of Structural Chemistry, 1992, Volume 33, Issue 6, PP 772-774.

[12] D. Bernal and R. H. Flower, J. of Chemical Physics Volume 1 (1933) No. 6 p 515.

of oxygen, called lone pair were placed in the other two corners. Because of repulsive forces of the two lone pairs, the two hydrogens are squeezed and lone pairs are separated as far as possible. The result is a bent structure for water with bond angle of 104.5° in gas phase.

This model was not universally accepted but was supported by many scientists because of supporting results obtained in various subsequent experiments and theoretical calculations.

In qualitative explanation of the tetrahedral structure, the three 2p orbitals and 2s orbital of oxygen (hydrogen type orbitals) are mixed to produce 4 orbitals called sp3 hybrid orbitals orientated toward the four corner of an imaginary tetrahedron. It should be pointed out that this mixing of atomic orbitals graphically is chemist visualization of approximate methods used for solving the Schrödinger equation.

As liquid, water is considered the molecule of life. Water can form four hydrogen bonds, two between its hydrogens and the oxygen atoms of two other water molecule, and two between its oxygen atom and hydrogens of two other water molecules. In liquid water, unlike ice the structure of liquid water is random and irregular, and the actual number of hydrogen bonds per liquid water molecule ranges from three to six, with an average of about 4.5. The necessity of maintaining a tetrahedral, hydrogen-bonded structure gives water an open loosely packed structure. The folding and self-assembly of proteins and DNA is attributed to ever changing network of hydrogen bond in liquid water.

Mark Gerstein and Michael Levitt used a super computer to create a model to explain how water affects the structure and movement of particular proteins[13]. 2013 Chemistry Noble awarded to Michael Levitt, Martin Karplus and Arieh Warshel as Noble Committee

---

[13] Simulating water and molecules of life, Mark Gerstein and Michael Levitt. Scientific America, November 1991, pages 101-105

put it for development of multi-scale models for complex chemical reactions.

C- Industrial applications

Hydrogen as a source of clean fuel

Hydrogen is being considered as clean fuel of the future. However, the main source of hydrogen is fossil fuel and natural gas which is not clean source of energy. Hydrogen is produced by processes called" Steam Reforming"

$$CH_4 + H_2O = CO + 3H_2$$

$$CO + H_2O = CO + H_2$$

Water electrolysis uses electricity from different source such as coal, fossil fuel, nuclear, or renewable sources of energy such as sun, wind, geothermal.

Water can be dissociated into diatomic molecules of hydrogen and (H_2) and oxygen O_2.

One mole of water produces one mole of hydrogen gas and half mole of oxygen gas. The Enthalpy of hydrogen molecule and oxygen molecule under the standard condition (298.15K and one bar pressure is taken to be zero, and entropy of hydrogen and oxygen are calculated with reference to zero entropy at 0K to be 130.68 J/K per mole and 205.14 J/K per mole respectively.

$$\Delta H^0_{298.15} = 1 \times \bar{H}^0_{298.15}(H_2, g) + \frac{1}{2} \times \bar{H}^0_{298.15}(O_2, g) - 1 \times \bar{H}^0_{298.15}(H_2O, l)$$

$$= 0 KJ + \tfrac{1}{2} \times 0 KJ - (-285.83 KJ) = 285.83 KJ$$

$$\Delta S^0_{298.15} = 1 \times \overline{S}^0_{298.15}(H_2, g) + \tfrac{1}{2} \times \overline{S}^0_{298.15}(O_2, g) - 1 \times \overline{S}^0_{298.15}(H_2O, l)$$

$$= 130.68 \text{ J/K} + \tfrac{1}{2} \times 205.14 \tfrac{J}{K} - 1 \times 69.91 J/K = 163.34 J/K$$

And

$$T\Delta S^0_{298.15} = 48.7 kJ$$

The entropy increases during electrolysis, this increase in entropy is compensated by decrease of entropy from the environment.

The energy for expansion from liquid water to hydrogen and oxygen gases is,

$$W = P_{external} \Delta V = (101.3 Pa) \times (1.5 moles) \times \left(22.4 \times 10^{-3} \frac{m^3}{mole}\right) \times \left(\frac{298K}{273K}\right)$$

$$= 3715 J$$

The ratio $\left(\frac{298K}{273K}\right)$ is introduce to calculate the volume of one mole gas at $298K$

Since $H = U + PV$, the change in energy is,

$$\Delta U^0_{298.15} = \Delta H^0_{298.15} - P\Delta V = 285.83 kJ - 3.72 k = 282.11 kJ$$ The amount of energy supplied by the battery is the change in Gibbs free energy.

Electrical work is defined as:

Electrical work = charge × potential

Therefore,

$$W_{elec} = n \times F \times E$$

Where, n is the mole number of electrons, E is the electromotive force of the cell and F is Faraday number which is equal to $96485.3365\ Coulomb\ mole^{-1}$. Since $J = coulom.volt$, then unit of W_{elec} is $J.mol^{-1}$

The work done on the electrochemical cell

$$-W_{elec} = -n \times F \times E$$

Maximum work besides mechanical work is given by equation 3-18,

$$\Delta G = W'$$, therefoe

$$\Delta G^0_{cell} = -nE^0_{cell}F$$

The amount of energy that should be supplied ideally (reversible process) is the change in Gibbs free energy,

$$\Delta G^0_{298.15} = \Delta H^0_{298.15} - T\Delta S^0_{298.15} = 285.83kJ - 48.7kJ = 237.1kJ$$

Recently a new technology called PlasmaArcFlow™ has been developed by R.M. Santilli[14] for the production of relatively clean

---

[14] R. M. Santilli, J. Foundation of hardonic Chemistry. With application to new clean energies and fuels (Kluwer Academic Publisher, Boston- Dordrecht- London, 2001

combustible gases called magnegases™. H_2, CO are the main components of magnegas with minor amounts of CO_2, H_2O, O_2 and atoms of H, O and C as well as radicals such as OH. These gases are being bonded to each other by strong magnetic fields from toroidal polarization of the valance and other electrons orbits.

Hydrogen Fuel Cells

Fuel cell is a device that converts the energy of a chemical reaction such as oxidation of hydrogen, natural gas and alcohols (methanol) with oxygen or other oxidizing agents. The cell consists of an anode, a cathode and electrolyte. In hydrogen fuel cell hydrogen is purged to the anode where it loses its electron and migrates through electrolyte to the cathode. The electron via external circuit is transferred to the cathode, where both, electron and hydrogen ion react with oxygen to produce water.

The first fuel cells were invented in 1838. One century later NASA used fuel cells to generate power in their space program. Commercial fuel cells are used for backup power and fuel cell vehicles such as forklifts, automobiles, buses and other vehicles.

Hydrogen as combustion fuel

The development of new car engines and hypersonic airplanes using hydrogen as fuel requires detailed and accurate information about the kinetic of reaction between hydrogen and oxygen. The reaction has explosion limits which depend on pressure and temperature. Conventionally the dependence of explosion limits on temperature and pressure loosely is divided into three regions. The original work on the reaction of hydrogen and oxygen by Hinshelwood goes back to 1920s[15]. Since then the reaction mechanism has been extensively investigated.

In modern approaches to hydrogen- oxygen combustion the goal is to reduce ignition delay and ignite the mixture at lower temperature. Since electronically excited atoms and molecules react much faster than non excited ones, then one approach will be to excite oxygen molecule to its lowest singlet state by electrical discharge at low pressure ($P = 10-20\,Torr$). The energy required for such excitation to lower singlet state requires $0.98\,eV$ ($94.555 kJ/mol$) energy and to vibrational excited state requires $0.193\,eV$ ($18.622 kJ/mol$) energy. Where, dissociation by collision requires $5.1\,eV$ ($492.1\,kJ/mol$) energy. In addition excitation to first singlet state may decrease the ignition temperature, reduce the induction time and improve the combustion of $O_2 - H_2$ mixture[16].

Ionization of water

Water coexist with hydroxide ion, H^-, and hydronium in, H_3O^+

$$HOH + HOH \rightleftharpoons OH^- + H_3O^+$$

---

[15] C.H. Gibson C. N. Hinshelwood, The homogenous reaction between hydrogen and oxygen, Proceedings Royal Society of London, A119: 591-606 (1928); A. H. Wilbourn and C. N. Hinshelwood, The mechanism of hydrogen – oxygen reaction I, The third explosion limit, Proceedings Royal Society of London, A18: 353-369 (1946).

[16] On combustion intensification mechanisms in the case of electrical-discharge-excited oxygen molecules Starik A M, Lukhovitskii B I, Naumov V V and Titova N S 2007 *Tech. Phys.* **52** 1281–90; On the influence of electronically excited oxygen molecules on combustion of hydrogen–oxygen mixture, V V Smirnov, O M Stelmakh, V I Fabelinsky, D N Kozlov, A M Starik and N S Titova A M Prokhorov General Physics Institute, Russian Academy of Sciences,Vavilov str. 38, 119991 Moscow, Russia

This is an equilibrium reaction and the equilibrium constant is given by

$$K_W' = \frac{[H_3O^+][OH^-]}{[H_2O]}$$

Since $[H_2O]$ is constant equal to 1000/18 = 55.56 M, and remains constant, then

$$K_w = [H_3O^+][OH^-] = 10^{-14} \text{ at 298 K}$$

Potentiometry, calorimetry and conductivity are the three main experimental techniques that have been used for measuring the ionization constant of water over wide ranges of temperature, potentiometry being one of the most precise techniques.

pK_w is defined as $-\log K_w = 14$ at 298 K.

For pure and neutral water $[H_3O^+] = [OH^-] = 10^{-7}$ at 298 K, by a similar definition $pH = -\log[H_3O^+]$ and $pOH = -\log[OH^-]$. Therefore, for neutral water at 298 K, $pH = pOH = 7$. As any equilibrium constant K_w depends on temperature.

Strength of an acid or base in aqueous solutions is determined by concentration of H_3O^+ or OH^-. The general formula for acid solution in water is

$$AH + H_2O \rightleftharpoons A^- + H_3O^+$$

The above definition of acid is in accord by Svante Arrhenius definition which defines acid as a substance that dissociates in water to produce H^+ and today definition of producing hydronium H_3O^+ ion. This definition is also in accord with Bronsted-Lowry of acid and base (1923). According to this definition acid is a substance that donates proton H^+ to a base, and base is a substance that accepts proton. Then AH is an acid and H_2O is a base and reaction produces a conjugate base A^- and a conjugate acid H_3O^+. In the reverse reaction H_3O^+ is an acid, AH is conjugate acid, A^- is a base and H_2O is a conjugate base.

In Arrhenius definition, base is a substance that produces OH^- ion in water. NH_3, amines and amides do not contain OH^- ion. The basicity is explained by $NH_3 + H_2O \rightleftharpoons NH_4^+ + OH^-$ reaction.

In Arrhenius concept of acid and base in the neutralization process salt and water are formed, that is H_3O^+ and OH^- form water and ionized salt. The Bronsted-Lowry definition dose not refers to the formation of water and salt, but to the formation of conjugate acid and conjugate base. The concept of neutralization is absent in their definition.

There are more general definition of acid and bases, that extends the definition of acid and base to other solvents, Lewis definition (1923), -Lux-Flood definition (1939), Usanovich definition (1938) and Pearson definition of hard-soft acid and base (1984).

Supercritical Water

The properties of water changes above 374°C and 218 atmposher, the density decreases and so does dielectric constant and strong electrolyte behaves as weak electrolyte. Hydrogen bonding becomes week, as the result supercritical water dissolves nonpolar compounds and loses its power for dissolving polar substances. On the basis of spectroscopic studies and computer modeling it is found that number of hydrogen bonds at supercritical conditions is about one third of hydrogen bonds at normal conditions[17].

Supercritical water is highly corrosive, and therefore it is necessary to use corrosion resistance alloys, or reactors lined with corrosion resistance alloy.

Because of its ability to dissolves non-polar compounds, thus it could be used as a safe solvent in the production of organic compounds. Supercritical water (SCW) is called green solvent.[18]

---

[17] Hydrogen Bonding in Supercritical Water. 2. Computer Simulations, G. Kalinichev, and J. D. Bass, *J. Phys. Chem. A* 1997, *101,* 9720-9727.

[18] Supercritical Water: A Green Solvent: Properties and Uses, Yizhak Marcus, John Wiley & Sons, Inc. 2012

www.ingramcontent.com/pod-product-compliance
Lightning Source LLC
Chambersburg PA
CBHW030857180526
45163CB00004B/1608